Advances in Optical Surface Texture Metrology

IOP Series in Emerging Technologies in Optics and Photonics

Series Editor

R Barry Johnson a Senior Research Professor at Alabama A&M University, has been involved for over 50 years in lens design, optical systems design, electro-optical systems engineering, and photonics. He has been a faculty member at three academic institutions engaged in optics education and research, employed by a number of companies, and provided consulting services.

Dr Johnson is an IOP Fellow, SPIE Fellow and Life Member, OSA Fellow, and was the 1987 President of SPIE. He serves on the editorial board of Infrared Physics & Technology and Advances in Optical Technologies. Dr Johnson has been awarded many patents, has published numerous papers and several books and book chapters, and was awarded the 2012 OSA/SPIE Joseph W Goodman Book Writing Award for Lens Design Fundamentals, Second Edition. He is a perennial co-chair of the annual SPIE Current Developments in Lens Design and Optical Engineering Conference.

Foreword

Until the 1960s, the field of optics was primarily concentrated in the classical areas of photography, cameras, binoculars, telescopes, spectrometers, colorimeters, radiometers, etc. In the late 1960s, optics began to blossom with the advent of new types of infrared detectors, liquid crystal displays (LCD), light emitting diodes (LED), charge coupled devices (CCD), lasers, holography, fiber optics, new optical materials, advances in optical and mechanical fabrication, new optical design programs, and many more technologies. With the development of the LED, LCD, CCD and other electo-optical devices, the term 'photonics' came into vogue in the 1980s to describe the science of using light in development of new technologies and the performance of a myriad of applications. Today, optics and photonics are truly pervasive throughout society and new technologies are continuing to emerge. The objective of this series is to provide students, researchers, and those who enjoy self-teaching with a wide-ranging collection of books that each focus on a relevant topic in technologies and application of optics and photonics. These books will provide knowledge to prepare the reader to be better able to participate in these exciting areas now and in the future. The title of this series is Emerging Technologies in Optics and Photonics where 'emerging' is taken to mean 'coming into existence,' 'coming into maturity,' and 'coming into prominence.' IOP Publishing and I hope that you find this Series of significant value to you and your career.

Advances in Optical Surface Texture Metrology

Edited by
Richard Leach
University of Nottingham, Nottingham, UK

IOP Publishing, Bristol, UK

ISBN 978-0-7503-2528-8 (ebook)
ISBN 978-0-7503-2526-4 (print)
ISBN 978-0-7503-2529-5 (myPrint)
ISBN 978-0-7503-2527-1 (mobi)

DOI 10.1088/978-0-7503-2528-8

Version: 20201201

IOP ebooks

British Library Cataloguing-in-Publication Data: A catalogue record for this book is available from the British Library.

Published by IOP Publishing, wholly owned by The Institute of Physics, London

IOP Publishing, Temple Circus, Temple Way, Bristol, BS1 6HG, UK

US Office: IOP Publishing, Inc., 190 North Independence Mall West, Suite 601, Philadelphia, PA 19106, USA

Contents

Preface

The book will cover the latest advances in the development of optical surface texture measuring instruments. Latest advances discussed will include the drive towards faster instruments for in-process applications, the ability to measure highly complex surfaces (in, for example, additive manufacturing) and advances in the use of artificial intelligence to enhance data analysis.

The book will mainly focus on manufacturing and precision engineering. The surface of an object is often the most critical feature when considering manufacturing tolerances, assembly with other parts and ultimately the functionality of the object. In addition, the range and complexity of surfaces being manufactured has recently increased significantly. Surface texture measurement in manufacturing is often carried out using contact stylus instrument, which is slow, requires contact, and produces small data sets (or minimal surface coverage). To increase throughput and allow the adoption of digital approaches, industry needs to use faster optical (non-contact) methods that produce dense point clouds. However, optical methods can be labour-intensive and expensive, especially when high-accuracy dense object coverage is required. Optical methods are also behind contact techniques in terms of a measurement traceability infrastructure, although the catch-up is well underway.

Due to the demands for more-accurate, faster and more automated optical surface measurement technologies, there have been many recent advances in terms of research, standardisation and industrial adoption. And, along with many other engineering disciplines, the world of optical metrology is starting to embrace the world of artificial intelligence. This book aims to update the reader in the latest advances in optical surface texture metrology.

Editor biography

Richard Leach

Richard is currently a Professor in Metrology at the University of Nottingham and prior to this spent 25 years at the National Physical Laboratory. He has been researching and lecturing on surface metrology for over 30 years. He is on the Council of the European Society of Precision Engineering and Nanotechnology, the Board of the American Society of Precision Engineering and several international standards committees. He is the European Editor-in-Chief for Precision Engineering and has over 500 publications including eight textbooks. He is a Fellow of the Institute of Physics, the Institution of Engineering & Technology, Higher Education Authority, the Institute of Measurement & Control, the International Society of Nanomanufacturing and the International Academy of Production Engineering (CIRP). He is a visiting professor at Loughborough University and the Harbin Institute of Technology.

List of contributors

Roger Artigas
Sensofar Tech SL, Terrassa, Spain
Roger Artigas has been working since 1997 at the Centre for Sensors, Instruments and Systems Development of the Universitat Politecnica de Catalunya as an optical engineer researcher. In 2001, he founded the company Sensofar Tech SL specialised in the design, manufacturing and commercialisation of optical measuring instruments. Since 2005 he has been part of the ISO TC 213 WG 16 to develop the ISO 25178 standard applied to the field of the instrumentation developed at Sensofar. He currently holds the position of President and CTO of Sensofar.

Reinhard Danzl
Bruker Alicona, Raaba, Austria
Reinhard Danzl is the head of software development at Bruker Alicona, which he joined in 2004. His main research interests are in the field of optical 3D metrology in general and the development of algorithms for focus variation and vertical focus probing in particular.

Zhengchun Du
Shanghai Jiao Tong University, Shanghai, People's Republic of China
Zhengchun Du is Associate Professor at Shanghai Jiao Tong University. He has worked as a Research Fellow at Shanghai Jiao Tong University and visiting scholar at University of California, San Diego and Georgia Tech. Since 2000, his primary research focus has been on the measurement, modelling and compensation of machine tool errors including geometric errors, thermally induced errors, cutting force induced errors. His research interests also include in-situ workpiece measurement on the machine tool and data-driven manufacturing process monitoring.

Xiaobing Feng
Shanghai Jiao Tong University, Shanghai, People's Republic of China
Xiaobing Feng is Assistant Professor at Shanghai Jiao Tong University. From 2014 to 2018, he was involved in several European research projects in the capacity as a Research Fellow at the University of Nottingham, focusing on manufacturing metrology. His research interests include surface texture measurement and characterisation, performance evaluation of surface and dimension measuring instruments, in-situ form and surface measurement on the machine tool and metrology assisted intelligent manufacturing.

Franz Helmli
Bruker Alicona, Raaba, Austria
Franz Helmli joined Bruker Alicona in 2001 and in 2003 became the head of the research and development department. Since 2006 has also concentrated on standardisation in ÖNORM, ISO TC 213 and VDI.

Mingyu Liu
University of Nottingham, Nottingham, UK
Mingyu Liu is a Research Fellow of the Manufacturing Metrology Team at the University of Nottingham. His research direction is on measurement and characterisation of precision freeform and structured surfaces using optical methods. His current research interest focuses on the development of light scattering and machine learning methods to characterise precision and additive manufactured surfaces.

Claudia Repitsch
Bruker Alicona, Raaba, Austria
Claudia Repitsch is an employee in the research and development department of Bruker Alicona. After working as a software developer in a company producing video slot machines, she was an EU-project researcher (iClass, ELeGI) at the University of Graz, and joined Alicona in 2008. She participates in the development of a quality management system according to ISO 9001 and other standards.

Rong Su
University of Nottingham, Nottingham, UK
Rong Su joined the University of Nottingham as a Research Fellow within the Manufacturing Metrology Team in 2015 following a short-term period working at the National Physical Laboratory in the UK. His expertise includes theoretical modelling, calibration and development of 3D imaging and light scattering instruments for surface and dimensional metrology. He is a member of the editorial board of Light: Advanced Manufacturing and Nanomanufacturing and Metrology, and a scientific committee member of several professional societies.

Wahyudin P Syam
University of Nottingham, Nottingham, UK
Wahyudin P Syam is a Research Fellow at the University of Nottingham within the Manufacturing Metrology Team and Nottingham Advanced Robotics Laboratory. His main research is in geometrical and surface texture measurements including calibration, performance verification and instrument development. Currently, he is working the development of an optical measuring system and framework to improve the accuracy of industrial robots. In addition, he is working on machine learning, particularly probabilistic approaches for measurement applications.

Jianguo Yang
Shanghai Jiao Tong University, Shanghai, People's Republic of China
Jianguo Yang is Professor at Shanghai Jiao Tong University. He was a visiting scholar at University of Michigan from 1995 to 1997. His research topics have mainly focused on the real-time compensation of machining errors in various types of CNC machine tools; from theoretical error modeling to the implementation of error compensation in synchronisation with the CNC control system. His research interests also include comprehensive modeling and compensation of

geometric and thermally induced errors, and parametric error compensation for ethernet distributed machine tool networks.

Kerstin Zangl
Bruker Alicona, Raaba, Austria
Kerstin Zangl is an employee in the research and development department of Bruker Alicona since 2011. She is a software developer and focuses on several research topics concerning the focus variation technology and the participation in the development of standards.

IOP Publishing

Advances in Optical Surface Texture Metrology

Richard Leach

Chapter 1

Terms and definitions

Richard Leach

In this chapter, some common terms and definitions are presented and discussed. Terms and definitions relating to metrology in general are given along with those specific to surface texture measurement and characterisation. Throughout the chapter, the relevant international standards are referred to and referenced. The concept of amplitude–wavelength space is briefly introduced as a way to compare the performance of surface measuring instruments.

1.1 Introduction

One of the most important aspects of a modern product is its surfaces; both shape and fine-scale topography are critical when considering tolerances, assembly and ultimately functionality (Leach 2014, Zhang *et al* 2015). Surfaces are highly sensitive to changes in a manufacturing process and can be diverse and complex due to various functional specifications and the characteristics of manufacturing processes. In tribology, surface interactions influence the friction, wear and lifetime of a component. It is estimated that surface effects cause 10% of manufactured parts to fail, which has a significant knock-on effect on a country's GDP (Leach 2014). In fluid dynamics surfaces determine how fluids flow and they affect, for example, aerodynamic lift, therefore influencing aircraft fuel consumption. Biomimetic surfaces can be engineered to significantly affect functions such as hydrophobicity, colour and hardness (Malshe *et al* 2018), structured surfaces are applied for food packaging applications (Karkantonis *et al* 2020) and there has been recent interest in how Covid-19 adheres to different surfaces (Gray 2020).

Over the past 100 years there has been significant effort devoted to the development of instruments and methods for surface measurement and characterisation. Whilst the contact stylus instrument has been the workhorse for surface measurement in industry, its non-contact nature, the need for areal surface measurement and the push for ever higher measurement speeds has resulted in an explosion of optical measurement technologies over the last thirty years. To address these optical technologies, a book was published in 2011 that summarised the working principles of the most popular optical instruments (Leach 2011). However, there has been significant progress since 2011 and this new book summarises that progress.

This chapter presents and discusses some common terms and definitions used throughout the book. Chapters 2 through 4 present the latest advances in coherence scanning interferometry, focus variation microscopy and imaging confocal microscopy, respectively, and chapter 5 reviews methods that do not require a vertical scan through focus. Chapter 6 reviews scattering techniques, along with methods for analysis using machine learning, and chapter 7 reviews the latest advances in in-process optical measurement technologies.

1.2 General metrology terms and definitions

(Adapted in part from Leach *et al* (2015).)

There are a number of terms relating to the field of metrology that need to be discussed briefly—these terms can also be used when reading all the other chapters. Any of these terms are used almost indistinguishably in practice, which can often lead to confusion when specifying instruments. The terms used in the chapter are taken from the latest version of the BIPM *International Vocabulary of Metrology* (BIPM *et al* 2012).

Traceability. The concept of traceability is one of the most fundamental in metrology and is the basis upon which all measurements can be claimed to be accurate. Traceability is defined as follows:

> Traceability is the property of the result of a measurement whereby it can be related to stated references, usually national or international standards, through a documented unbroken chain of comparisons all having stated uncertainties.

It is important to note the last part of the definition of traceability that states *all having stated uncertainties* (see the definition below). This is an essential part of traceability as it is impossible to usefully compare, and hence calibrate, instruments without a statement of uncertainty. Uncertainty and traceability are inseparable (Haitjema 2013). Traceability applied to surface texture metrology is discussed elsewhere (Leach *et al* 2015).

Calibration is defined as follows:

> Operation that, under specified conditions, in a first step establishes a relation between the quantity values with measurement uncertainties provided by measurement standards and corresponding indications with associated measurement uncertainties and, in a second step, uses this information to establish a relation for obtaining a measurement result from an indication.

In simpler terms, calibration is a comparison between two measurements, one of which is a reference or standard value, and the other of which is being tested. Again, note the use of the term uncertainty in the formal definition of calibration. Commonly the term calibration is misused, which has led to confusion in understanding the aim of the calibration process. The frequent misuse of the calibration term is when it is confused with adjustment.

Adjustment is defined as follows:

> A set of operations carried out on a measuring system so that it provides prescribed indications corresponding to given values of a quantity to be measured

The adjustment process physically changes some parameters of a metrological tool (it can be a mechanical adjustment or it could be the result of changing the value of a software constant) to provide an indication that is closer to a known value. The adjustment process does not provide information about measurement uncertainty. Similar results could be obtained by correcting the measurement result using the results from a calibration certificate. A meaningful measurement result can be presented without adjustment, but it must have an associated uncertainty.

An example of adjustment of a surface texture measuring instrument is when the manufacturer adjusts the field of view of an optical instrument by determining the flatness deviation (Leach *et al* 2015) and, from this, applies a correction (for example, Ekberg *et al* 2017). Many modern instruments have some form of software error correction applied. The adjustment cannot account for the uncertainty associated with the measurement result; it only uses a value from the range of possible values that are within the limits given by the measurement uncertainty. After this adjustment, the measurement of a surface will provide a different result. The basic difference between calibration and adjustment is also illustrated by the requirement in ISO 17025 (ISO 2005) that an instrument should be calibrated before and after adjustment.

Verification is defined as follows:

> Provision of objective evidence that a given item fulfils specified requirements.

A verification test is designed to check whether a particular instrument attribute meets its specification. Verification, therefore, does not necessarily imply that measurement uncertainty is part of the test, but usually some form of quantitative measure is required. Often, an assertion that an instrument is within specification assumes that the test result is inside the specification by at least a 'guard band', for example, the expanded uncertainty. There are currently no standard verification methods in surface texture metrology is although this will be covered by ISO/CD 25178 part 700 (ISO/DIS 2020) and discussed briefly elsewhere (Leach *et al* 2015).

Measurement uncertainty is defined as follows:

> A non-negative parameter characterising the dispersion of the quantity values being attributed to a measurand, based on the information used.

It is generally accepted that the basis of uncertainty estimation is given in the *Guide to the Expression of Uncertainty in Measurement* (GUM) in its most recent version: JCGM 100 (JCGM 2008a). Supplement 1 was also published in 2008, entitled *Propagation of Distributions using a Monte Carlo Method* (JCGM 2008b). Uncertainty estimation for surface texture metrology is discussed elsewhere (Leach *et al* 2015).

1.3 Surface terms and definitions

(Adapted in part from Leach *et al* (2015).)

Surface metrology is the science and application of surface measurements. However, there is some controversy over how to define the instruments that effectively measure surface texture. As defined below, surface texture is something that is calculated from measured topography data, therefore in ISO 25178 part 600 (ISO 2019) the instruments are referred to as 'surface topography measuring'. Surface texture measuring instruments are, therefore, not defined.

Typically, a surface topography measuring instrument has the ability to move a probing system (sensor) that detects the location of the surface on an object and the capability to determine spatial coordinate values on the surface relative to a reference coordinate system.

The following definitions are used in this book:

Surface form—the underlying shape of a part (Leach 2014) or fit to a measured surface (ISO 2010a).

Surface topography—all the surface features treated as a continuum of spatial wavelengths (Leach 2014). Basically, here surface topography is everything that makes up the geometry of the object's surface, i.e. it is the surface form plus the surface texture.

Surface texture—the geometrical irregularities present at a surface. Surface texture does not include those geometrical irregularities contributing to the form or shape of the surface (Leach 2015). Simply put: surface texture is what is left of the surface topography once the surface form has been removed. Figure 1.1 illustrates the relationship between surface topography, form and texture.

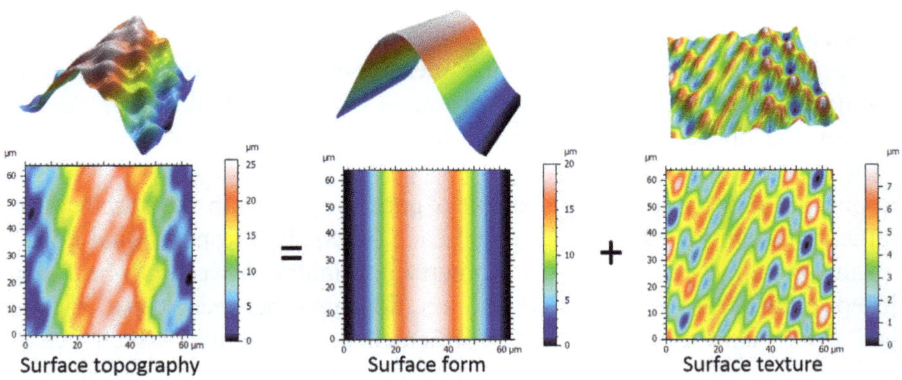

Figure 1.1. Different surface types.

Form and texture make up topography, but it is not always obvious where to draw the distinction between them. Some modern complex surfaces can also blur the lines between what is considered texture and what is considered form, as the fine-scale structure is often of the same order in size as the features that make up the shape of the object. Despite this blurred distinction, surface texture and form metrology tend to be very different fields, including different specification standards and ways in which measurement traceability is demonstrated.

Surface profile measurement is the measurement of a line across the surface that can be represented mathematically as a height function with lateral displacement $z(x)$. *Areal surface texture* measurement is the measurement of an area on the surface that can be represented mathematically as a height function with displacement across a plane $z(x, y)$. Surface texture characterisation, both profile and areal, is presented in detail elsewhere (Whitehouse 2010, Leach 2013, 2014).

There is a range of instrumentation for measuring surface topography, and this book will consider the latest advances. A previous book on the subject of measurement using optical methods covers all the basic techniques and the current book is effectively the next 'chapter'. It is advisable that the reader is familiar with Leach (2011) before reading the current book.

ISO 25178 part 6 (ISO 2010b) defines three classes of methods for surface texture measuring instruments:

Line profiling method. A surface topography method that produces a 2D graph or profile of the surface irregularities as measurement data that may be represented mathematically as a height function $z(x)$. Examples given in ISO 25178 part 6 include: stylus instruments, phase shifting interferometry, circular interferometric profiling and optical differential profiling.

Areal topography method. A surface measurement method that produces a topographical image of the surface that may be represented mathematically as a height function $z(x, y)$. Often $z(x, y)$ is developed by juxtaposing a set of parallel profiles. Examples cited in ISO 25178 part 6 include: stylus instruments, phase shifting interferometry, coherence scanning interferometry, confocal microscopy, confocal chromatic microscopy, structured light projection, focus variation microscopy, digital holography microscopy, angle resolved SEM, SEM stereoscopy, scanning tunnelling microscopy, atomic force microscopy, optical differential and point autofocus profiling.

Area-integrating method. A surface measurement method that measures a representative area of a surface and produces numerical results that depend on area-integrated properties of the surface texture. An example given in ISO 25178 part 6 is total integrated scatter.

Amplitude–wavelength (AW) space is a good way to map the specifications of instruments (Stedman 1987, Jones and Leach 2008). Each operational constraint (for example, range, resolution, tip geometry and lateral wavelength limit) can be modelled and parameterised, and the relationships between these parameters derived. The relationships are best represented as inequalities which define the area of operation of the instrument. A useful way to visualise these inequalities is to construct a space where the constraining parameters form the axes. The constraint

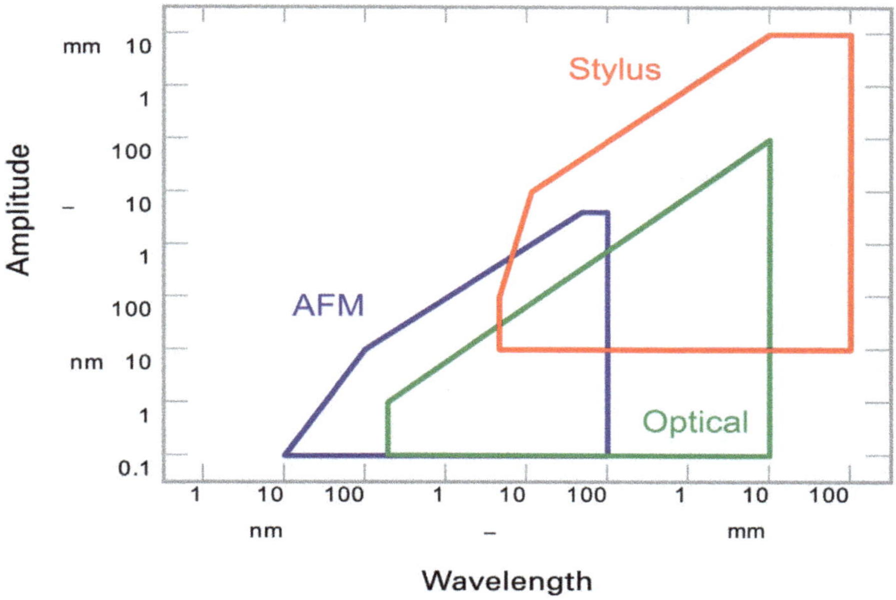

Figure 1.2. Amplitude–wavelength map for three common instrument types.

relationships (inequalities) can be plotted to construct a polygon. This shape defines the viable operating region of the instrument. Constraints that are linear in a given space must form a flat plane across that space, solutions on one side of which are valid. Such a plane can only form a side of a convex polyhedron containing the viable solutions. Figure 1.2 shows an AW space plot for three common instruments.

References

BIPM, IEC, IFCC, ILAC, ISO, IUPAC, IUPAP and OIML 2012 *International Vocabulary of Metrology—Basic and General Concepts and Associated Terms—JCGM 200* (Sèvres: Bureau International des Poids et Mesures)

Ekberg P, Su R and Leach R K 2017 High-precision lateral distortion measurement and correction in coherence scanning interferometry using an arbitrary surface *Opt. Express* **25** 18703–12

Gray R 2020 Covid-19: How long does the coronavirus last on surfaces? *BBC Future* https://bbc.com/future/article/20200317-covid-19-how-long-does-the-coronavirus-last-on-surfaces

Haitjema H 2013 Measurement uncertainty *CIRP Encyclopedia of Production Engineering* ed L Laperrière and G Reinhart (Berlin: Springer)

ISO 17025 2005 *General Requirements for the Competence of Testing and Calibration Laboratories* (Geneva: International Organization of Standardization)

ISO 10110 part 8 2010a *Optics and Photonics—Preparation of Drawings for Optical Elements and Systems—Part 8: Surface texture; roughness and waviness* (Geneva: International Organization for Standardization)

ISO 25178 part 6 2010b *Geometrical Product Specification (GPS)—Surface Texture: Areal—Part 6: Classification of methods for measuring surface texture* (Geneva: International Organization for Standardization)

ISO 25178 part 600 2019 *Geometrical Product Specifications (GPS)—Surface Texture: Areal—Part 600: Metrological characteristics for areal topography measuring methods* (Geneva: International Organization of Standardization)

ISO/DIS 25178 part 700 2020 *Geometrical Product Specifications (GPS)—Surface Texture: Areal—Part 700: Calibration of surface topography measuring instrument* (Geneva: International Organization of Standardization)

JCGM 2008a *Evaluation of Measurement Data—Guide to the Expression of Uncertainty in Measurement (GUM 1995 with minor corrections)—JCGM 100* (Sèvres: BIPM)

JCGM 2008b *Evaluation of Measurement Data—Supplement 1 to the Guide to the Expression of Uncertainty in Measurement—Propagation of Distributions using a Monte Carlo Method—JCGM 101* (Sèvres: BIPM)

Jones C W and Leach R K 2008 Adding a dynamic aspect to amplitude–wavelength space *Meas. Sci. Technol.* **19** 055105

Karkantonis T, Gaddam A, See T L, Joshi S S and Dimov S 2020 Femtosecond laser-induced sub-micron and multi-scale topographies for durable lubricant impregnated surfaces for food packaging applications *Surf. Coat. Technol.* **399** 126166

Leach R K 2011 *Optical Measurement of Surface Topography* (Berlin: Springer)

Leach R K 2013 *Characterisation of Areal Surface Texture* (Berlin: Springer)

Leach R K 2014 *Fundamental Principles of Engineering Nanometrology* 2nd edn (Amsterdam: Elsevier)

Leach R K 2015 Surface texture *CIRP Encyclopaedia of Production Engineering* ed L Laperrière and G Reinhart (Berlin: Springer)

Leach R K, Giusca C L, Haitjema H, Evans C and Jiang X 2015 Calibration and verification of areal surface texture measuring instruments *Ann. CIRP* **64** 797–813

Malshe A P, Bapat S, Rajurkar K P and Haitjema H 2018 Bio-inspired textures for functional applications *Ann. CIRP* **67** 627–50

Stedman M 1987 Basis for comparing the performance of surface-measuring machines *Precis. Eng.* **9** 149–52

Whitehouse D J 2010 *Handbook of Surface and Nanometrology* 2nd edn (Boca Raton, FL: CRC Press)

Zhang S J, To S, Wang S J and Zhu Z W 2015 A review of surface roughness generation in ultra-precision machining *Int. J. Mach. Tools Manuf.* **91** 76–95

IOP Publishing

Advances in Optical Surface Texture Metrology

Richard Leach

Chapter 2

Coherence scanning interferometry

Rong Su

Coherence scanning interferometry is a well-established and widely used technique for areal surface topography measurement. Technological progress has been continuously made to face challenges from the advanced manufacturing sector and materials research. This chapter reviews the latest advances in coherence scanning interferometry in the past two decades, covering aspects of applications, system designs, surface reconstruction algorithms, theoretical modelling and methods for calibration and adjustment.

2.1 Introduction

Coherence scanning interferometry (CSI) is one of the most accurate techniques for surface topography measurement of industrial materials. CSI is a type of reflection-mode interference microscopy in which a three-dimensional (3D) image can be acquired as a stack of two-dimensional (2D) images by scanning the object along the optical axis through the focal plane of the imaging objective lens (figure 2.1).

In addition to the mechanical scanning operation, a major difference between CSI and conventional wide-field microscopy is that CSI employs an interferometric objective lens which usually has a built-in Mirau or Michelson interferometer. Michelson objectives are mostly used for low numerical aperture (NA), low-magnification systems (10× and lower), while Mirau objectives are more useful at higher NA and higher magnification (10× to 100×). CSI systems can also be constructed using a Linnik interferometer configuration, where two ordinary microscopic objectives are used in the two paths of the interferometer, respectively. This configuration is sometimes used for high-magnification systems, when a larger working distance than with a Mirau objective is desired. Optical microscopes with incoherent illumination can resolve small features in the lateral direction with a resolution up to $\lambda/2A_N$ according to Abbe's diffraction limit (where λ is the wavelength of the incident light and A_N is the NA of the optical system). A lens with small NA can be used for the benefit of a large field of view (FOV). Although

doi:10.1088/978-0-7503-2528-8ch2

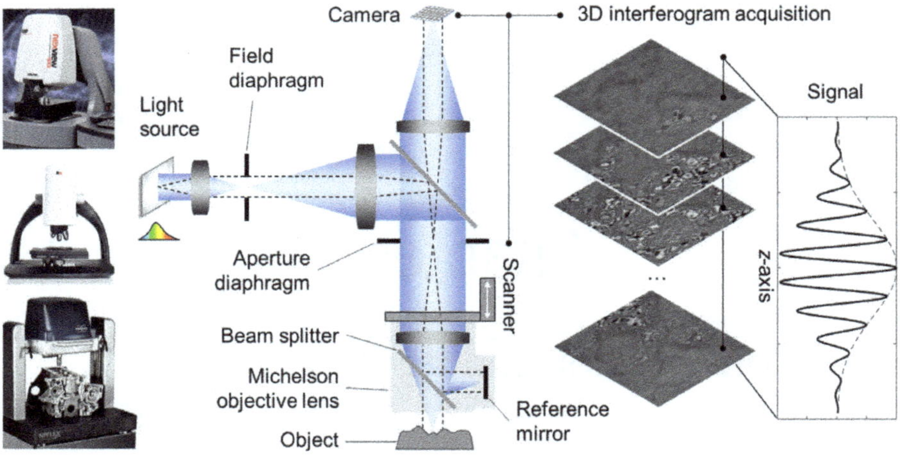

Figure 2.1. Overview of CSI technique. Left: commercial systems (image courtesy of Zygo Corporation, Sensofar and Bruker). Middle: schema of a typical CSI system. Right: 3D image stack acquired through axial scanning and typical CSI signal recorded at a camera pixel.

the lateral resolution will be reduced, the measurement sensitivity in the axial direction is retained. This feature is not afforded by non-interferometric microscopic imaging systems.

In addition, CSI is distinguishable from laser phase-shifting interferometry (PSI) as it uses a broadband light source, for example, a light-emitting diode (LED). The light source usually has a central wavelength in the visible region and a spectral width of a few tens of nanometres. The source is why CSI has also been known as 'scanning white-light interferometry', although the radiation emitted from the source does not often appear white.

When a surface is scanned through the focus of a CSI system, interference fringes will form only within a (axial) window of a few micrometres around the surface, corresponding to the zero optical path length difference of the interferometer. The axial size of this window (i.e. the coherence envelope) correlates with the coherence length which is determined by the degree of temporal and spatial coherence of the illumination. This phenomenon is also known as 'low-coherence interference' and provides high-accuracy optical sectioning capability. When an objective lens with a small NA is used (the illumination pupil is determined by this NA), the degree of spatial coherence is high, the low-coherence interferogram is achieved mainly by the low temporal coherence of the spectrally broadband source. When a lens with large NA is used, the degree of coherence is determined by both the NA and the light source spectrum (this will be discussed in section 2.4).

In CSI, the surface topography is reconstructed by analysing the 3D interferogram. Assuming the broadband source has a Gaussian spectrum in wavenumber (inverse wavelength), the interference fringe will be modulated by a coherence envelope which is also Gaussian. This envelope can be used to determine the surface height, or to estimate the zero-order fringe to eliminate the well-known 2π ambiguity that is inherent in laser interferometry and PSI. As the fringe order is obtained, the phase of the fringe can be used to determine the surface height without applying complex phase unwrapping. This unique feature allows CSI to measure a wide range of surfaces, from flat or smooth surfaces to stepped, discontinuous or rough surfaces.

Several reviews on the working principle and applications of CSI can be found in Schmit *et al* (2007), Hariharan (2007), Damian *et al* (2009), Bauer *et al* (2013) and de Groot (2011, 2015a, 2015b, 2019). This chapter will focus on the advances in this technique in the past two decades. The reader is expected to be familiar with the basic working principle of CSI.

Section 2.2 summarises the applications of CSI in different industries and research areas. Section 2.3 reviews the recent technical progress in CSI in the development of both hardware and surface reconstruction algorithms. Section 2.4 reviews advances in the theoretical modelling and understanding of image formation in CSI. Finally, calibration and adjustment methods are discussed in section 2.5.

Related techniques, such as low-coherence interferometry for distance sensing, wavelength (or frequency) scanning interferometry (see chapter 5), digital holographic microscopy, optical coherence tomography/microscopy and interferometric confocal microscopy are not included in this chapter.

2.2 Applications

CSI has been applied in a broad range of applications for measurements of diverse and complex surfaces in various manufacturing sectors, such as semiconductors, optics, automotive, aerospace, medical devices and energy.

2.2.1 Surface finish characterisation

Surface texture parameters can be calculated from the measured topography. Some examples of using CSI are the surface texture evaluation of the panels of large radio antenna for astronomical observations (Chinellato *et al* 2010), the texture of 316-stainless steel of standard masses (Laopornpichayanuwat *et al* 2012), the surface quality control of fuel injection systems (Sachs and Stanzel 2014), the texture analysis of rocks for mineral identification and studies in rock movement on different surfaces (Mukhtar *et al* 2016), the effects of texture on the potential surface colonisation by micro-organisms for cement paste surfaces (Apedo *et al* 2016), topography measurement of turbine blade surfaces (Zou *et al* 2016), and surface form and subsurface damage of ground glass (Sergeeva *et al* 2010, Bae *et al* 2017).

Super-polished surfaces can also be measured by using hybrid data acquisition incorporating sinusoidally modulated phase shifting. This technique increases the signal-to-noise ratio (SNR) and reduces the noise density to the 0.1 nm $Hz^{-1/2}$ level (Fay *et al* 2014a).

In the biomedical field, the effects of surface topography on the cytotoxicity of Ge–Sb–Se chalcogenide glass optical fibres has been studied (Mabwa *et al* 2020). CSI has also been used to characterise the topography of specially designed and laser-processed surfaces for applications in the food industry, where the correlation between surface texture parameters and laser processing parameters has been investigated, as well as the influence of texture on cellular adhesion and the formation of biofilms on surfaces (Lazzini *et al* 2017). By analysing the surface gradient information from CSI-measured topography, the orientation distribution of crystalline grains can be rapidly characterised, providing a result comparable to the standard electron backscatter diffraction method (Speidel *et al* 2018). Together with feedback analysis and finite element modelling, a closed-loop laser texturing process was enabled by integrating a miniaturised CSI sensor head inside the laser texturing machine, parallelising both texturing and surface measurement (Bermudez *et al* 2019).

2.2.2 Measurement of highly sloped surfaces

For a mirror-like surface that only reflects light in the specular direction, the highest surface slope that can be measured by CSI is determined by the NA of the objective lens. Light specularly reflected by a surface that has a slope larger than $\arcsin(A_N)$ will fall outside the acceptance cone of the objective lens.

Advances in the CSI technique now allow for measurement of non-specular surfaces with slopes beyond the traditional NA acceptance cone (Fay *et al* 2014b, Thomas *et al* 2020a). This capability is attributed to the capture of diffuse and back-scattered light from the microstructures found on the surface slopes. The SNR of the back-scattered signal can be enhanced by using high-dynamic range measurement that is performed through the alternation of light levels during a measurement, or by increasing the sampling frequency of the interference fringe (known as 'signal oversampling'). Details of these techniques will be discussed in section 2.3.

2.2.3 Additive manufactured surfaces

Surface topography can be considered as the fingerprint of an additive manufacturing process. By studying the surface features, such as weld tracks, weld ripples, attached particles and surface recesses in a metal powder-bed fusion (PBF) process, it is possible to extract useful information to optimise the additive manufacturing process, and consequently improve product quality and reduce energy and material consumption (Townsend *et al* 2016, Leach *et al* 2019). However, additive manufactured surfaces usually have high levels of texture and contain a large number of features with high slope angles and loose particles. Accurate measurement is a challenging task for contact and non-contact methods. Some example CSI measurements of additively manufactured surfaces are shown in figure 2.2.

Measurements of additively manufactured metallic surfaces made using various optical and non-optical technologies were compared, including imaging confocal and focus variation microscopy, CSI and x-ray computed tomography. The differences were investigated to increase the understanding of the behaviour and performance of areal topography measurement for additive manufactured surfaces (Senin *et al* 2017).

With recent technical advances, such as high-dynamic range measurement and signal oversampling techniques, state-of-the-art commercial CSI systems are capable of providing high-quality measurements of additive manufactured plastic and

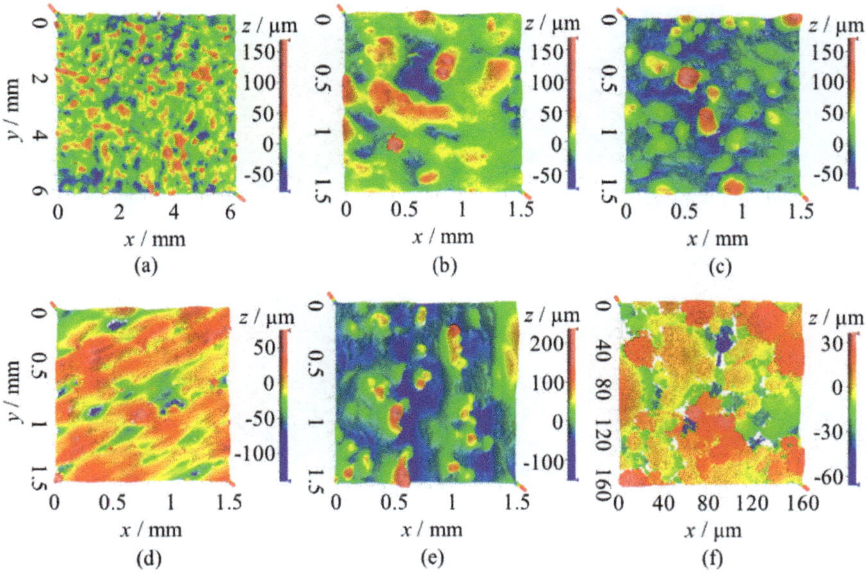

Figure 2.2. Additively manufactured surfaces measured by CSI. (a) and (b) Laser PBF Al–Si–10Mg cube top surfaces, (c) laser PBF Al–Si–10Mg cube side surface, (d) laser PBF Ti–6Al–4V cube top surface, (e) electron beam PBF Ti–6Al–4V rectangular prism top surface and (f) laser PBF Ti–6Al–4V cube side surface (reprinted with permission from Gomez *et al* 2017, copyright SPIE).

metallic parts (Fay *et al* 2014c). The effects of these advanced measurement functions were investigated through a series of experiments on additive manufactured metal surfaces made of various materials and processes. Recommendations for measurement optimisation regarding data coverage, measurement area and measurement time were provided (Gomez *et al* 2017).

An empirical investigation into the evolution of the topography of an inkjet printed transparent polymer was conducted using CSI (Gomez *et al* 2020a). This work covered a variety of inkjet printed polymer structures with several geometries such as dots, lines, films and honeycombs. The CSI measurements provided an insight into how to control and optimise the quality of inkjet printed parts.

2.2.4 Film measurement

The fringe contrast of CSI depends on the refractive index contrast of two neighbouring media, and the instrument's optical sectioning capability is mainly determined by the coherence length of the source. It has been shown that CSI is suitable for the measurement of transparent or semi-transparent layers and films, for example, films on silicon wafers (Chang *et al* 2006), rough and translucent hydroxyapatite layers (Pecheva *et al* 2007), conducting oxides, thin film photovoltaics, carbon and ceramic coatings on silicon (Yoshino *et al* 2017, Yu and Mansfield 2015, Maniscalco *et al* 2014), multi-layered films (de Groot and Colonna de Lega 2008) and ceramic coatings on metallic surfaces with complex topography (Feng *et al* 2019), see figure 2.3.

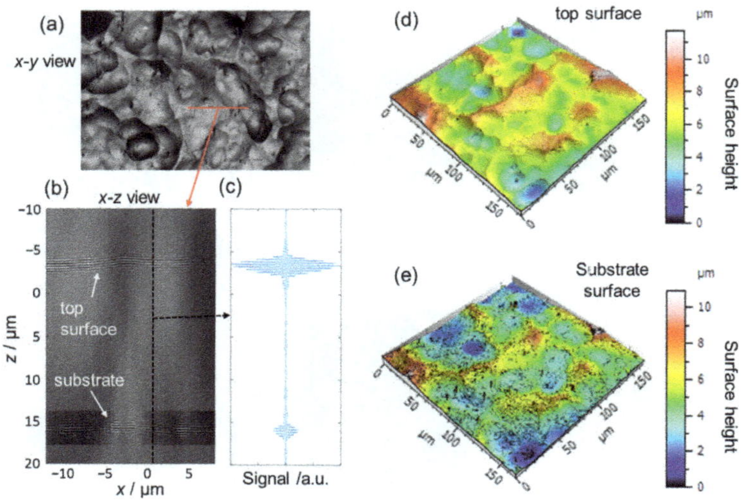

Figure 2.3. CSI measurement of a ceramic coating (Feng *et al* 2019). (a) Top-view intensity map of the measured area in the x–y plane, (b) an arbitrarily-chosen cross-sectional slice through the 3D CSI image, (c) 1D interferogram at a single point on the surface along the z-axis, and topography of (d) the top surface and (e) the substrate from which the coating thickness can be calculated.

The optical thickness of the film can be measured as the distance between the peaks of the two coherence envelopes if the film thickness is sufficiently large, i.e. larger than the coherence length of the illumination source. The physical thickness can be easily calculated if the refractive index of the film material is known.

If the film thickness is smaller than the coherence length, for example, from ~10 nm to ~2 μm, it becomes difficult to separate the surface and substrate of the film by searching for coherence peaks. In this case, model-based film analysis techniques can be used to measure film thickness down to 10 nm (de Groot and Colonna de Lega 2007, de Groot and Colonna de Lega 2008, Fay and Dresel 2017). This method is based on modelling of the interferometric signal and is followed by a library search. In the most basic implementations this approach requires *a priori* information of the complex refractive indices of the film and the substrate; however, multi-variable library fits can solve for the indices simultaneously.

Another approach to measure thin film with a thickness of tens of nanometres is the 'helical conjugate field' function method (Mansfield 2008). This function equates to a topographically defined helix modulated by the reflectance of the film. As such, it provides a 'signature' of the thin film structure so that, through optimisation, the thin film structure can be determined on a local scale. In order to use the helical conjugate field function method, it is necessary to provide *a priori* knowledge of the dispersive film index or use reference film structures with known properties.

2.2.5 Functional imaging

It has been shown that model-based methods can be extended to determine the spectral refractive index of a substrate or absorbing film. This capability has been demonstrated on oxide-on-silicon with/without gold coating (Fay and Dresel 2017), as well as on surfaces of silicon, gold and a gold/palladium alloy using silica and zirconia oxide thin films (Yoshino *et al* 2016). By using a sample of known reflectivity for calibration, it is possible to extract the spectral signature of the CSI system and to deduce the local reflectance spectra of a surface of interest. The reflectance of the surface can be calculated over the given wavelength range of the effective spectrum, which is defined as the source spectrum multiplied by the spectral response of the camera and the spectral transmissivity of the optical system (Claveau *et al* 2016). Through introduction of a datum plane, the refractive index distribution and thickness distribution can be measured simultaneously (Watanabe *et al* 2014).

It has been shown that a stroboscopic PSI can visualise surface acoustic wave propagation on a surface (Nakano *et al* 1997). CSI has also been used to measure oscillating surfaces based on the stroboscopic effect (de Groot 2006). By using a stroboscopically synchronised supercontinuum light source incorporated into a scanning low-coherence interferometer operating in the wavelength region of 1.1–1.7 μm, Hanhijärvi *et al* (2013) measured the surface of the hidden face of a thermally actuated oscillating 4 μm thick silicon microelectromechanical system bridge. Heikkinen *et al* (2013) built a stroboscopic CSI system operating with a 400–620 nm broadband 150 mW light source. The source combined a non-phosphor white LED with a cyan LED. Measurement of the surface topography of a capacitive micromachined ultrasonic transducer membrane operating at 2.72 MHz was demonstrated.

2.3 Technical advances

The technical advances of CSI in the past two decades are reviewed, covering topics on surface reconstruction algorithms, enhancement of noise performance and spatial bandwidth, new light sources and new system designs.

2.3.1 Surface reconstruction methods

The CSI interferogram recorded by each camera pixel along the axial scanning direction is modulated by the coherence envelope. The position of the envelope provides a first estimate of the surface height and fringe order, and then the fringe phase can be used to refine this estimate. Various methods have been developed for reconstructing surface topography from fringes. They are based on three major principles—envelope detection (Larkin 1996, Sandoz 1997), spatial frequency domain analysis (de Groot and Deck 1995, de Groot *et al* 2002), and correlation between the measured and reference fringes (Kino and Chim 1990). Harasaki *et al* (2000) developed a five-frame algorithm that determines both the best-focus frame position and the fractional phase from the best-focus frame of the interferogram acquired through vertical scanning to remove fringe-order ambiguity. Pavlíček and Michálek (2012) investigated the effects of noise on the uncertainty of the envelope detection method based on a Hilbert transformation. Gianto *et al* (2016) and Montgomery *et al* (2013) compared fringe analysis methods based on Hilbert and wavelet transforms, and introduced an algorithm using the Teager–Kaiser energy operator. Hamprecht (2005) introduced a Bayesian approach that unifies surface reconstruction and post outlier removal processes. Guo *et al* (2011) used the Carré phase shifting algorithm to extract the phase information from CSI signal. This method allows a flexible choice of the equi-spaced phase steps of the actuator.

Many algorithms have been developed with some modifications of the basic approaches to improve the surface height measurement for tilted and curved surfaces, or at sharp edges where errors due to phase jumps occur. A few examples are given elsewhere (Ghim and Davies 2012, Ma *et al* 2011, Vo *et al* 2017, Huang *et al* 2020). The improvements in the measurement results were achieved through digital filtering processes for some specific samples.

2.3.2 Enhancement of the signal-to-noise ratio

Signal oversampling and high-dynamic range methods are implemented into modern CSI systems to enhance the SNR (Fay *et al* 2014a, Gomez *et al* 2017). Usually, four sample points are used for sampling a fringe (along the axial direction), i.e. four camera frames per fringe, corresponding to a sampling distance of one eighth of the central wavelength. Increasing the sampling frequency is referred to as 'signal oversampling'. When using this technique, the number of camera frames per fringe is increased by sampling the fringe at smaller phase increments and the scan speed is reduced. As more sample points represent a signal, surface topography can be reconstructed with lower uncertainty, according to the central limit theorem (Gomez *et al* 2020b). The high-dynamic range technique allows multiple exposures with different light levels or exposure times to be collected in sequence, and a composite image is formed from the image data with the highest SNR, commonly gauged by the contrast between each pixel and its neighbours.

Signal oversampling methods may enhance the SNR for surfaces with low reflectance, high levels of texture or steep slope angles. The high-dynamic range of the light levels optimises the signal strength for a surface with a large variation in reflectance and/or slope. Both methods can significantly increase the number of valid data points for measurements of complex surfaces (Gomez *et al* 2017).

In addition, a method has been presented to reduce the effect of speckle on measurement uncertainty when measuring rough surfaces (Wiesner *et al* 2012). By sequentially changing the direction of the illumination, the camera captures several independent speckle patterns in sequence. From each pattern, the brightest speckles are selected to calculate the height map. The result showed an effective reduction of outliers.

2.3.3 Enhancement of measurement resolution and bandwidth

The lateral resolution in CSI is diffraction-limited. Based on modelling of the scattered field from a square grating with a 190 nm pitch and 80 nm lateral width using rigorous coupled wave analysis (RCWA), the fringe data of a CSI system (central wavelength 570 nm) was synthesised and a library of example signals was generated. Through a library search in a least-squares sense with the acquired signal data, the lateral critical dimension of this optically unresolved grating structure was measured with 1 nm sensitivity and the result agrees with that obtained using an atomic force microscope (de Groot *et al* 2008).

CSI systems usually have a FOV ranging from 100 μm to a few millimetres depending on the imaging optics. Olszak (2000) developed a white-light interferometer that scans the object laterally with a tilted coherence plane. This technique allows measurements at higher speeds and the measurement range can be increased without stitching in one direction. A similar approach has been reported elsewhere (Chen *et al* 2014).

Data fusion methods have been used to extend the spatial dynamic range of CSI. For enhancement of the lateral resolution when a low-magnification lens is used, the low-resolution surface topography data can be fused with corresponding high-resolution intensity data. The fusion method is based on an intrinsic image-based shape-from-shading algorithm (Wang *et al* 2018). For enhancement in the measurement range, multi-sensor data fusion approaches have been developed, taking advantage of the strengths of different techniques. A method for the fusion of photogrammetry and CSI data allows the photogrammetry data to be accurately scaled with reference to the CSI data, and in turn the exact locations of multiple high-resolution CSI measurements can be determined in the coordinate system defined by photogrammetry. The ultimate goal is to allow for high-accuracy 3D optical coordinate measurement and surface topography measurement simultaneously (Leach *et al* 2018).

2.3.4 Pupil-plane imaging in CSI

Imaging the pupil plane (i.e. back focal plane) of a high-NA microscope objective onto a camera provides the angle-resolved optical properties of surfaces at a single, focused measurement point. The white-light interference pattern collected at the pupil plane can be converted into ellipsometric information and allows the measurement of the thickness and refractive index of thin film reference surfaces. The information has also been used to create accurate 3D topography maps of complex object structures (Colonna de Lega and de Groot 2008).

Based on pupil-plane imaging, Ferreras Paz *et al* (2012) combined the Fourier scatterometry method with CSI to improve the depth sensitivity of the scatterometry. They demonstrated the measurement of sub-wavelength features of a silicon line grating by using a model-based reconstruction method to compare simulated and measured spectra (see figure 2.4).

2.3.5 Light sources

LEDs or incandescent lamps are the most common light sources for CSI. An alternative advanced light source is a femtosecond pulsed laser that generates a train of ultrafast laser pulses. The high-spatial coherence of the laser enables a large FOV with high fringe visibility and allows for a high-volume measurement of semiconductor chips, flat panel displays and photovoltaic cells (Oh and Kim 2005). By scanning the pulse repetition rate with direct reference to a rubidium atomic clock, step heights of approximately 67 μm are measured with a repeatability of 10 nm

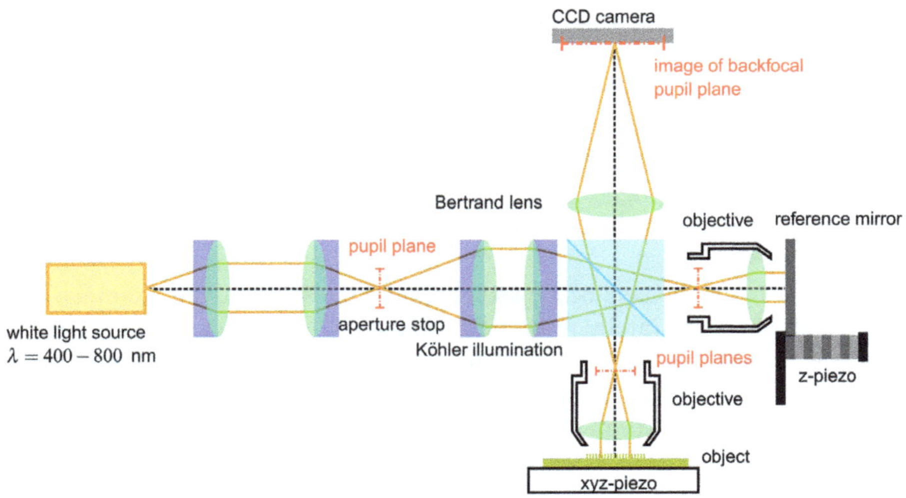

Figure 2.4. Schematic CSI set-up for pupil-plane imaging (reprinted with permission from Ferreras Paz *et al* 2012, copyright 2012 CIOMP).

(Joo *et al* 2013). Using an erbium-doped fibre femtosecond laser with ~30 μm temporal coherence length, a CSI system has been developed to measure rough silicon carbide surfaces (Lu *et al* 2018). A supercontinuum light source has been used in a custom-built CSI system (Kassamakov *et al* 2009). This light source is based on a Nd:YAG pumped micro-structured optical fibre and provides a high output intensity of 20–35 mW, which is beneficial when measuring low-reflectivity samples.

Systematic errors often occur in the measurement of highly sloped surfaces and sharp edges. A second LED emitting light at a different mean wavelength was used to reduce deviations between height values obtained from the coherence envelope and the phase (Niehues *et al* 2007).

2.3.6 System design

This section introduces new designs of interferometers, lenses and scanning methods that are related to CSI technique. A high-NA CSI system has been developed based on the principle of the geometric phase, which is close to achromatic in nature (Roy *et al* 2002). It has been shown that the 'batwing effect' or 2π phase jumps in CSI can be minimised by using the geometric phase-shifting technique (Roy *et al* 2009).

To measure super-smooth surfaces, such as hard disk substrates and super-polished optics, an interferometer configuration using a virtual reference surface has been designed to eliminate the mid- to high-spatial frequencies from the reference beam (Freischlad 2012). A residual systematic waviness error of less than 0.02 nm (root-mean-square) was demonstrated.

A tandem interferometer and a Michelson interferometer with achromatic polarising optics were built for absolute length and form measurements at a large working distance (>150 mm) (Ullmann *et al* 2015). The tandem interferometer measures the zero-point position of a white-light signature with a peak-to-peak difference of 154 nm under uncontrolled environmental conditions without thermal

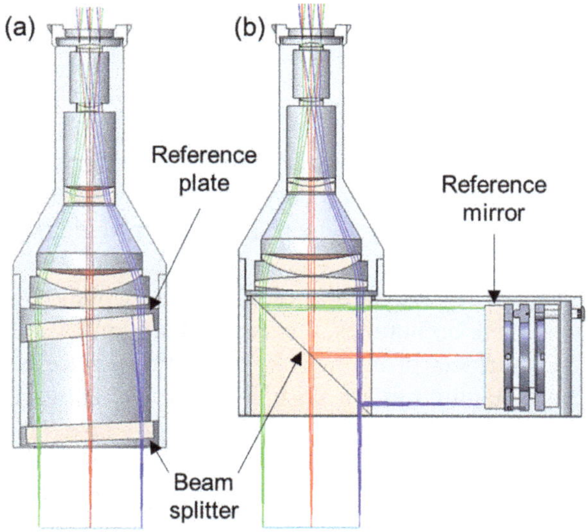

Figure 2.5. Interferometric objective lens design (image courtesy of Zygo Corporation). (a) The 'de Groot–Biegen' interferometer and (b) the traditional Michelson geometries for a 0.5× interferometric objective.

stabilisation. The white-light Michelson interferometer with polarising achromatic optics allows zero-point detection with a standard deviation of less than 15 nm.

New interferometric objectives have been developed to extend the range of applications for flexible microscope platforms to larger FOVs, including a turret-mountable 1.4× objective (see figure 2.5) and a dovetail mount 0.5× objective with a 34 mm × 34 mm FOV (de Groot and Biegen 2016). This type of objective is called the 'de Groot–Biegen' interferometer and comprises a beam-splitter plate and a partially transparent reference mirror arranged coaxially with the objective lens system. The coaxial plates are slightly tilted to direct unwanted reflections outside of the imaging pupil aperture, providing high-fringe contrast with spatially extended white-light illumination.

To measure dynamic samples, Dunsby *et al* (2003) developed a single-shot CSI that utilises spatially separated phase-stepped images and requires only one camera to achieve simultaneous acquisition of four phase-stepped images. By using a simultaneous phase sensor, Wiersma and Wyant (2013) developed a CSI system to produce repeatable measurements over an extended range. The prototype system can measure a 4.5 μm step height artefact in the presence of vibration amplitudes of 40 nm with a repeatability of 1.5 nm and a 400 nm vertical scanning step size. To enhance the measurement speed, Jeon *et al* (2019) developed a CSI system using a polarised CMOS camera based on the spatial phase-shifting technique. This system also allows a vertical scanning step size larger than that determined by the Nyquist sampling limit.

Conventional CSI requires mechanically scanning of the object relative to the measuring system to form an image stack. A CSI system has been developed in which the reference plane moves and the CSI uses an electrically focus-tuneable lens to ensure that the object is sharply imaged during the measurement procedure (Pavliček and Kučera 2019, Pavliček and Mikeska 2020).

2.4 Theoretical modelling of CSI

Understanding and modelling of image formation is important for advancing the CSI technique and is the foundation of the system design and surface reconstruction. A measurement model significantly influences the metrological capability of a CSI system.

Section 2.4.2 considers 2D image formation in the x–y plane that is orthogonal to the optical axis (i.e. z-axis). Sometimes, the term 2D imaging also refers to the image formation of a 1D surface profile of a prismatic surface in the x–z plane, corresponding to an imaging system with a slit pupil. However, the latter case will be referred to as a quasi-3D model as it can usually be extended to the full 3D imaging case by adding the other horizontal coordinate (see section 2.4.3).

2.4.1 One-dimensional signal modelling

The simplest model of CSI signal generation is based on the basic 1D low-coherence interferometry model. The contrast of interference fringes is determined by the optical path length difference between the beams reflected back from the sample and the reference mirror. In CSI, the fringe is modulated by the coherence envelope, determined by the light source spectrum. The 1D model does not consider the effects of light scattering by the sample surface and assumes that the phase of the light reflected from the sample is proportional to the surface height along the optical axis. However, the 1D model can incorporate high-NA and spectral effects by means of an incoherent superposition of ray bundles through the interferometer spanning a range of wavelengths, incident angles and pupil-plane coordinates (de Groot and Colonna de Lega 2004). The superposition can be carried out efficiently in the frequency domain, followed by a Fourier transform to generate the simulated interference signal. This method can also be useful for understanding the signal formation of thin film structures in CSI. A study of the effects of thin film on a test surface which includes the effects of multiple reflections has been provided by Roy *et al* (2005). The practical case of a dielectric film on a metallic substrate (an oxidised silicon surface) has been discussed.

2.4.2 Two-dimensional imaging model and transfer function

A 2D CSI model considers the effects of surface diffraction/scattering and the finite 2D pupil of the imaging system. As a type of interference microscope, the capability of CSI for resolving the fine structures of a surface obeys the Abbe theory of image formation, i.e. the scattered/diffracted light needs to be captured within the pupil to provide image contrast. Under the spatially shift-invariant condition, the imaging capability of a 2D imaging system can be characterised by its 2D transfer function (TF), specifically the 2D coherent transfer function (CTF, also known as the amplitude transfer function) for coherent illumination, which is the scaled pupil function, and the 2D optical transfer function (OTF) for incoherent illumination, which is the autocorrelation of the 2D CTF. The theory of TF in optical systems has been well-documented in, for example, Goodman (2017).

To study the batwing artefact near the sharp edge of a step height, a diffraction model has been developed by treating the step edge as two shifted apertures. The batwing artefact occurs when the step height is less than the coherence length of the light source (Harasaki and Wyant 2000). A quasi 2D model has been developed by Xie *et al* (2012), where a sequence of interference signals along the axial z-direction at each horizontal position of the surface was first formed using a 1D model, and then the 3D image stack was filtered by considering a 2D OTF. Surface scattering effects were not considered in this model.

A 2D elementary Fourier optics (EFO) model for topography measurement in interference microscopy has been described in detail by de Groot and Colonna de Lega (2020). This model is based on the small surface height approximation for the surface scattering problem. In the EFO model, the surface height variation is assumed to be much smaller than the depth of field, which is given by the wavelength divided by the square of the NA. In this approximation, the surface is represented by its mean plane and the phase of the total field on the plane can be considered proportional to the surface height. The 2D image formation is then modelled by propagation of the 2D angular spectrum of plane waves. The 2D CTF is used to filter the angular spectrum. To further simplify the EFO model, the concept of effective wavelength can be used which averages the effect of different illumination wave vectors with spatially extended illumination. In this case, the 2D angular spectrum corresponding to the effective wavelength is filtered by the 2D OTF of the imaging system. For surfaces that are flat relative to the depth of field, the 2D OTF can be used to approximately characterise the instrument transfer function (ITF), where the ITF is defined as the ratio of Fourier components for the measured and true topographies. Although subject to the small height approximation, the EFO model is useful for predicting the main features of CSI topography images for a wide class of surfaces, and balances the simplicity, accuracy and computational efficiency.

2.4.3 Three-dimensional imaging model and transfer function

An analysis of 3D image formation is of significance as CSI has been widely used to measure 3D surfaces with significantly large height deviations. 3D optical imaging techniques can be considered as 3D linear shift-invariant filtering operations (Coupland and Lobera 2008). The imaging process is characterised by the 3D TF of the imaging system and this concept has been widely used in the area of optical microscopic imaging (Gu 2000). The 3D CTF is determined by the generalised 3D pupil that has been introduced by McCutchen (1964).

The 3D model of an imaging system, based on the concept of the generalised 3D pupil, is not limited by the paraxial condition and can, therefore, be applied for systems with large NA. The 3D CSI imaging model combines 3D imaging theory with an appropriate surface scattering model. For example, under the Kirchhoff approximation (also known as the Kirchhoff theory for surface scattering, or the wave optics method), image formation in CSI can be characterised by its 3D surface transfer function (STF) which applies to the 'foil model' of the surface. At each x and y position, the foil model is a 1D Dirac delta function in the z-direction which is biased by the surface height at this position (Coupland *et al* 2013, Su *et al* 2020a). The linearisation of the imaging process is mainly restricted by the validity regime of the Kirchhoff approximation, of which the main condition is that the curvature at a surface point must be small (Beckmann and Spizzichino 1987).

The 3D STF is a complex-valued quantity (figure 2.6)—its magnitude determines the spatial frequency passband of the CSI, and the phase reveals the optical aberrations of the imaging system (Su *et al* 2020b). The 3D point spread function (PSF) and spatial resolution for topographic measurement can be easily derived from the 3D STF.

It is not necessary to use the Kirchhoff approximation for the surface scattering process. Any appropriate scattering model can be combined with the generalised 3D CSI imaging model, including exact solutions to the Maxwell's equations for the scattering problem (which will be introduced in section 2.4.4).

A quasi-3D CSI model based on the Kirchhoff approximation for scattering has been proposed for studying the signal formation in high-NA (0.9) CSI systems of a rectangular grating (Xie *et al* 2016, Lehmann *et al* 2018). The model described can only be applied for the 1D profile of a prismatic surface, and the incident and

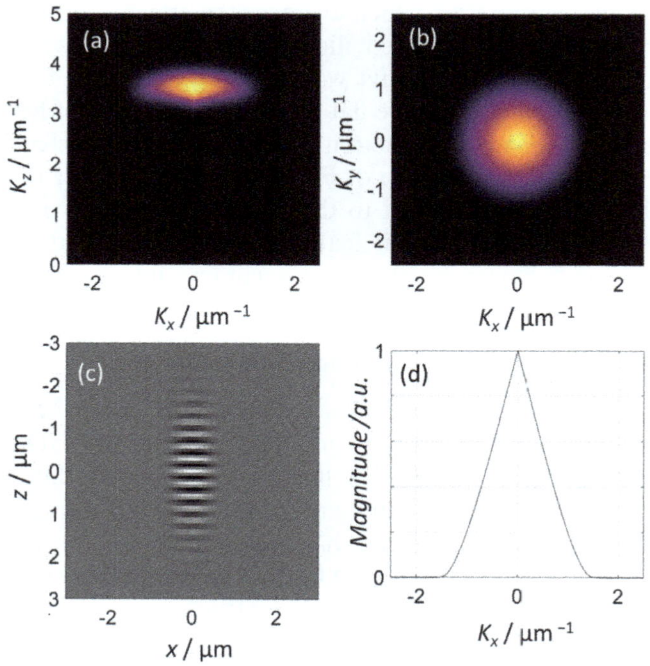

Figure 2.6. Theoretical 3D STF and PSF of a CSI system operating at 550 nm, spectral full width half maximum 60 nm and NA 0.4. $K_{x,y,z}$ are the spatial frequencies with respect to the x-, y- and z-directions. (a) Cross section of the 3D STF in the K_x–K_z plane, (b) the 2D in-pupil-plane STF calculated as a projection of the 3D STF into the K_x–K_y plane, (c) cross section of the 3D PSF and (d) a profile through the origin of (b). Note that the phase term of the 3D STF is zero for a diffraction-limited system.

scattered waves are limited in the plane of incidence, i.e. the x–z plane. In principle, this model should yield the same result as the 3D model described in Coupland *et al* (2013) and Su *et al* (2020a) but for a CSI system with a slit pupil.

Surfaces featuring sharp edges, fine pitches and large angles of incidence violate the validity regime of the Kirchhoff approximation. Therefore, CSI models based on this approximation should be used with caution when studying surfaces such as rectangular gratings. Fortunately, if the amplitude of the grating is small compared to the wavelength, the Kirchhoff approximation is closely related to the small surface height approximation as used in the 2D EFO model (Su *et al* 2020a).

Another CSI model introduced by Xie *et al* (2016) and Lehmann *et al* (2018) is called the Richards–Wolf model. This model is based on the Debye diffraction integral which describes the diffraction by the aperture of an imaging system and is closely related to the concept of the generalised 3D pupil. The Debye diffraction integral is not relevant to surface scattering/diffraction. In fact, the small surface height approximation was used in these works to predict the scattered field from the rectangular grating.

In summary, in the linear (single-scatter) regime, the most commonly used 2D/3D CSI models are all based on the Kirchhoff and small surface height approximations.

2.4.4 Image formation beyond the linear regime

For complex surfaces that contain sharp edges and points, vee-grooves, high aspect-ratio holes or re-entrant features, multiple surface scattering cannot be neglected. Therefore, rigorous scattering models are required, which typically use numerical techniques to solve Maxwell's equations exactly. A small number of CSI models based on rigorous scattering exist.

As introduced in section 2.3.3, CSI fringe data can be simulated using RCWA to predict the scattered field from a square grating with a 190 nm pitch (de Groot *et al* 2008). RCWA approaches are usually only appropriate for periodic structures (Moharam and Gaylord 1983). A boundary element method (BEM) has been developed to predict scattering for arbitrarily complex surfaces (Simonsen 2010). In contrast to the finite element method's (FEM's) volume discretisation, BEM solves linear partial differential equations along only the surface and, therefore, is more computationally efficient. Given a specific incident plane wave, the BEM model predicts the angularly resolved far field scattering. The scattering amplitude is then used to generate the CSI data in the spatial frequency domain through a holographic recording process. Finally, the interferogram corresponding to a specific angle of incidence and wavelength is formed through a Fourier transform of the recorded data in the spatial frequency domain (figure 2.7). In this model, the spatial domain interferograms resulting from different angles of incidence and wavelengths need to be generated iteratively and synthesised incoherently to give the final CSI image (Thomas *et al* 2020b).

A simpler rigorous approach uses FEM to first generate the near-field scattered field on a line (for a slit aperture) or a 2D plane (for a circular aperture) in the

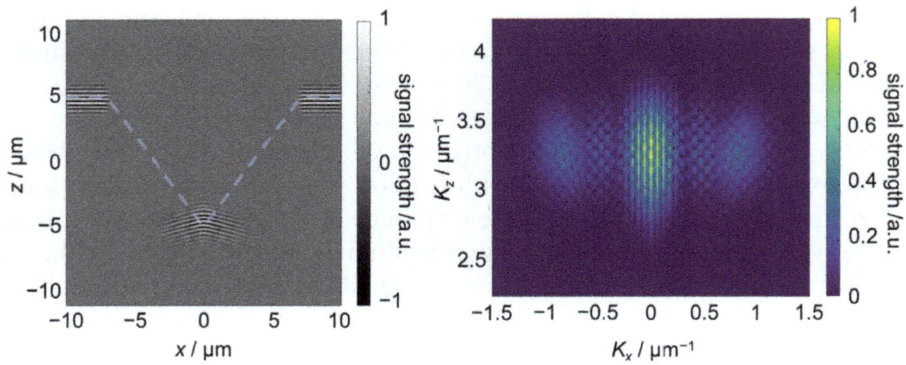

Figure 2.7. Synthesised CSI interferogram (a) for a vee-groove with the dihedral angle of 70° and the depth of 10 μm based on a rigorous surface scattering solver and (b) the magnitude of the interferogram in the spatial frequency domain (Thomas *et al* 2020b). $K_{x,z}$ are the spatial frequencies with respect to *x*- and *z*-directions.

vicinity of the surface. Then, the field at the image plane can be calculated by propagation of the (1D/2D) angular spectrum with consideration of the finite pupil (Bischoff *et al* 2020).

Because rigorous methods are computationally expensive and time consuming, the existing rigorous CSI models were based on RCWA/BEM/FEM algorithms that restrict the incident and scatter wave vectors to the plane of incidence, i.e. the *x*–*z* plane. However, there is no major theoretical barrier and a 3D rigorous solver for the scattering problem has been recently developed (Coupland and Nikolaev 2019).

2.5 Calibration and error correction

Measurement accuracy is of primary importance for a metrological instrument. General calibration and estimation of uncertainty for surface topography measurement are summarised (Leach *et al* 2015). In this section, recent progress on calibration and error correction techniques for CSI will be summarised.

2.5.1 Noise

The instrument noise in CSI has often been confused by the term 'axial/vertical resolution' which is a widely quoted and misunderstood performance specification for commercial surface topography measuring equipment (de Groot 2017). Axial resolution is somewhat meaningless for surface topography measurement as there are no adjacent points to be resolved in the axial direction if only the outer surface of an object is of interest. An exception is the measurement of films, coatings or subsurface features, in which cases axial resolution is an indicator of the optical sectioning capability.

Noise is a random error source and may be quantified through repeatability tests. Noise evaluation methods for CSI are based on averaging or subtraction to isolate the measurement noise from the sample surface topography (Haitjema and Morel 2005, Giusca *et al* 2012a). Noise density, which incorporates together with noise the effects of the data acquisition time and the number of independent surface

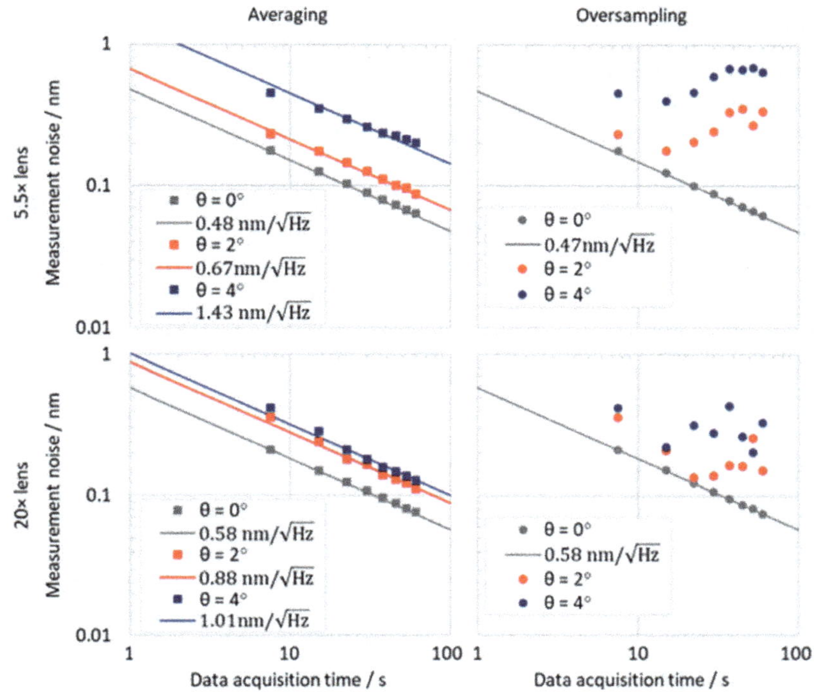

Figure 2.8. Measurement noise in CSI reduces with a square root dependency on the total data acquisition time. An optical flat is measured at different tilt angles θ. The noise density values for corresponding tilt angles are given in the legend (reprinted with permission from Gomez *et al* 2020b, copyright Springer Nature).

topography data points, has been introduced to CSI to allow more meaningful comparisons (de Groot and DiSciacca 2018).

Instrument noise corresponds to measurement noise under the best possible conditions. The instrument noise of a CSI system is often evaluated by measuring a levelled optical flat. A major cause of the instrument noise is camera shot noise which follows a square root dependency on the total data acquisition time. Pavliček and Hýbl (2012) have studied the theoretical lower limit of the instrument noise caused by shot noise using the Cramer–Rao inequality. However, both experimental and simulation results show that the measurement noise levels rise when increasing the tilt angle of the optical flat and in the presence of environmental vibration (Gomez *et al* 2020b), with similar result elsewhere (Liu *et al* 2015). The topography averaging method is effective for reducing the noise regardless of surface tilt (figure 2.8). The signal oversampling method, introduced to improve the SNR in individual data acquisitions, can also reduce the final measurement noise; however, in the presence of vibration, oversampling is effective in reducing the measurement noise only when the surface height variation or tilt are smaller than the depth of field (Gomez *et al* 2020b). Alternatively, vibration caused by the piezoelectric actuator in a CSI system can be reduced using the input shaping technique, where a second command signal is designed to cancel the vibration of the first command for the actuator (Mun *et al* 2015, Song *et al* 2018).

2.5.2 Linearity and amplification

The traditional method for determining the amplification coefficient and linearity of the z-axis of a CSI system is by measuring calibrated step height artefacts. To extend the calibrated range, multiple overlapped measurements of a step height artefact can be used (Giusca *et al* 2012b).

An alternative to the traditional method for establishing the amplification coefficient (height scale) of a CSI system has been proposed (de Groot and Beverage 2015). This method links wavelength standards (obtained by using a spectrometer to provide traceability to the 546.074 nm 198 Hg line) to a calibration of the properties of the optical path length scanning mechanism of the interferometer. In many cases, this provides a lower uncertainty compared to calibration using step height artefacts, with direct traceability to the metre.

2.5.3 Lateral distortion

Lateral optical distortion is present in most CSI systems and may cause field-dependent systematic errors in surface measurement. The traditional method of determining the lateral distortion is by using areal cross grating artefacts (Giusca *et al* 2012b, Henning *et al* 2013). However, the calibration is often limited by the accuracy of the physical artefacts.

A method based on image correlation and a self-calibration technique allows for calibration of the lateral distortion in a CSI system by using an arbitrary surface that contains some surface features (Ekberg *et al* 2017). This technique was demonstrated using a coin surface, and nanometre level precision was achieved over an FOV of 1.5 mm. However, an absolute linear scale is needed to make the scale traceable.

2.5.4 Optical aberrations

Optical aberrations, such as the spherical aberration and dispersion (Pförtner and Schwider 2001, Colonna de Lega 2004, Lehmann 2010, Lehmann *et al* 2014, Hovis *et al* 2019) as well as other high order aberrations (Su *et al* 2020b), degrade the measurement accuracy in CSI, see figure 2.9. In addition, the effects of defocus on the 3D STF in CSI has been studied (Su *et al* 2018).

Most aberration correction methods attempt to adjust the measured surface topography through digital data processing, such as phase unwrapping or low-pass filtering methods. Such methods may cause additional errors for some surfaces and may degrade the lateral resolution. Some recent research demonstrated an aberration correction method that acts on the raw fringe data without any digital data processing (Su *et al* 2020b). This method requires accurate information about the 3D STF of the CSI system (see section 2.5.7).

2.5.5 Multiple materials

In a phase measuring instrument, phase changes on reflection depend on the optical properties of the materials (Doi *et al* 1997, Harasaki *et al* 2001). The phase changes can cause height measurement errors of the order of 10–100 nm in CSI. By

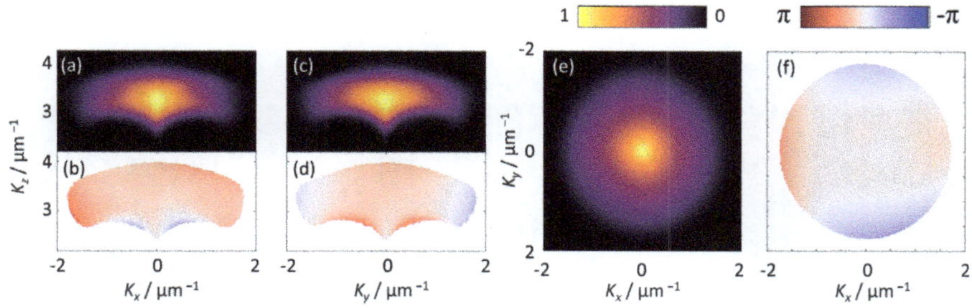

Figure 2.9. Experimentally measured 3D STF of a commercial CSI system operating at approximately 560 nm, full width half maximum spectral width ~100 nm and NA 0.55 (Su *et al* 2020b). (a) and (c) Vertical cross sections of the magnitude of the 3D STF, (b) and (d) the corresponding phase, and (e) and (f) the magnitude and phase of the 2D in-pupil-plane STF calculated as a projection of the 3D STF into the K_x–K_y plane. Note that the phase term is non-zero in a real system as compared to the aberration-free case as shown in figure 2.6.

measuring the optical properties of the material, this error can be compensated (Palodhi *et al* 2011, Raid *et al* 2015). A model-based CSI can provide phase change on reflection-corrected topography for any combination of dissimilar materials with known visible-spectrum refractive indices, including metals (Fay and Dresel 2017).

2.5.6 Steeply sloped surfaces

Engineered functional surfaces often feature varying slope angles on the macro- and micro-scales. For a mirror-like surface, the highest surface slope angle that can be measured is arcsin(A_N), with respect to the *x*- or *y*-axis; known as the 'NA limit'. A surface with a larger slope angle cannot be reliably measured. In reality, many surfaces are not mirror-like and produce a distribution of scatter when illuminated. Capture of the backscatter allows a CSI system to measure the form of a tilted surface outside the NA limit. However, significant errors have been observed for texture measurements (figure 2.10) and it is still unclear how the instrument responds to textures on a tilted surface (Thomas *et al* 2020a).

One approach to potentially improve the capability of CSI to measure large surface slope angles has been proposed (Coupland and Lobera 2010). The concept was demonstrated by simulations. CSI interferograms were synthesised for the case of vee-grooves, step artefacts and re-entrant features using a FEM-based scattering model, where the effects of multiple scattering were considered. The improved capability for measuring large surface slopes was achieved by first tilting the sample and subsequently by using an iterative FEM model to provide improved estimates of the illuminating conditions.

In addition, it has been understood that when measuring a tilted flat, the fringe recorded at a pixel has a different spacing and coherence envelope compared to the fringe for a level flat (Sheppard and Larkin 1995, Lehmann *et al* 2019).

2.5.7 Instrument transfer function and resolution

From a user perspective, the ITF (defined as the ratio of Fourier components for the measured and true topography) is important as it characterises the instrument's

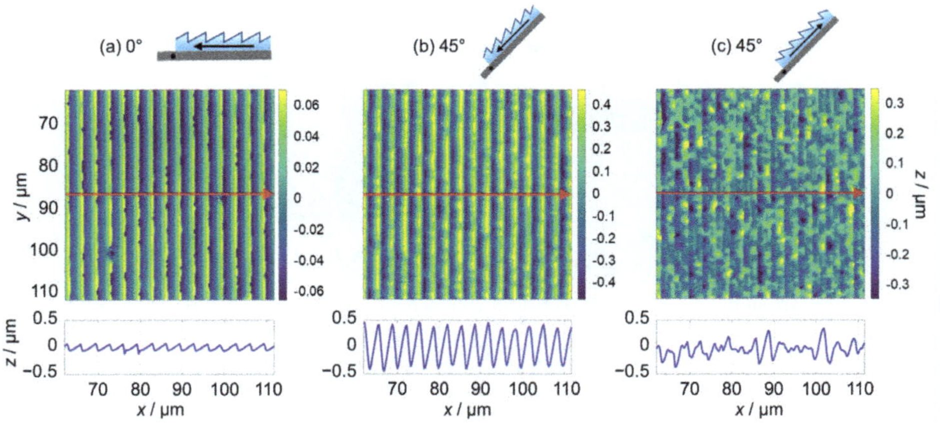

Figure 2.10. Topography results after a best-fit plane removal from a CSI measurement of a blazed reflective diffraction grating (with a nominal peak-to-valley amplitude of 200 nm and a frequency of 300 lines/mm) with tilting arrangements of (a) 0°, (b) 45° downwards and (c) 45° upwards, using a Mirau objective lens with a 0.55 NA.

response to the spatial frequencies of a surface (Colonna de Lega and de Groot 2012). The ITF corresponds to an evaluation of the complete system of which the instrument response is assumed to be linear in spatial frequency within defined limits. The optics is only part of a CSI system but the most important part. Thus, the OTF or similar TF for the optics as part of a model can be used for predicting the ITF of a CSI system (de Groot *et al* 2019).

In the limit of small surface heights, the ITF of a CSI system can be approximated by the 2D OTF (Colonna de Lega and de Groot 2012). Under the Kirchhoff approximation, the 3D STF of a CSI system provides the instrument's imaging response to a surface in the 3D spatial frequency domain (Coupland *et al* 2013, Su *et al* 2020a, 2020b), and the ITF can be approximated by the 2D projection of the 3D STF, although the validity of the ITF is still subject to the limit of small surface heights.

The 3D STF of a CSI system can be experimentally determined by measuring the surface of a precision sphere of which the diameter is known with high accuracy and sphericity error is negligible. The diameter of the sphere should be much larger than a wavelength and smaller than the FOV. Under the linear systems theory of CSI (Coupland *et al* 2013), the 3D STF is calculated as the ratio between the 3D interferogram and the foil model of the spherical cap in the spatial frequency domain (figure 2.9). The foil model is defined by a 1D Dirac delta function $\delta[z - S(x, y)]$, where $S(x, y)$ is the surface height map in the Cartesian coordinate system. The method for characterisation and correction of the 3D STF for a CSI system was first demonstrated in Mandal *et al* (2014). The effects and tolerance of the sphere form errors on the accuracy of the 3D STF were studied by Su *et al* (2017a, 2017b). The characterisation and correction method has been later verified with freeform surfaces (Su *et al* 2020b).

From the 3D STF, the resolution of a CSI system can be easily determined according to different criteria. Traditionally, the lateral resolution is determined using a physical artefact with periodic structures, for example, a star-shaped material measure (Eifler *et al* 2019). The resolution is defined as the (tangential) spatial period of the star-shaped material measure at which the height response of an instrument falls to 50% (Giusca and Leach 2013).

Acknowledgements

Special thank gives to Professor Peter de Groot from Zygo Corporation for his comments on this chapter. The colour maps in figures 2.6 and 2.9 are extracted from Kovesi (2015).

References

Apedo K L, Montgomery P, Serres N, Fond C and Feugeas F 2016 Geometrical roughness analysis of cement paste surfaces using coherence scanning interferometry and confocal microscopy *Mater. Charact.* **118** 212–24

Bae E, Kim Y, Park S and Kim S-W 2017 Large-aperture ground glass surface profile measurement using coherence scanning interferometry *Opt. Express* **25** 1106

Bauer W, Weber M and Chanbai S 2013 White light interferometry *Encyclopedia of Tribology* (Boston, MA: Springer US) pp 4115–27

Beckmann P and Spizzichino A 1987 *The Scattering of Electromagnetic Waves from Rough Surfaces* (Norwood, MA: Artech House)

Bermudez C, Martínez P, Cadevall C and Artigas R 2019 High-speed, roll to roll coherence scanning interferometry in a laser texturing process *Proc. 19th EUSPEN* pp 176–7

Bischoff J, Pahl T, Lehmann P and Manske E 2020 Model-based dimensional optical metrology *Proc. SPIE* **11352** 113520P

Chang S P, Xie T B and Sun Y L 2006 Measurement of transparent film using vertical scanning white-light interferometry *J. Phys. Conf. Ser.* **48** 1063–7

Chen L-C, Le M-T and Lin Y-S 2014 3-D micro surface profilometry employing novel Mirau-based lateral scanning interferometry *Meas. Sci. Technol.* **25** 094004

Chinellato S, Pernechele C, Carmignato S and Manzan F 2010 Surface measurements of radio antenna panels with white-light interferometry *Proc. SPIE* **7739** 77392T

Claveau R, Montgomery P C, Flury M and Montaner D 2016 Local reflectance spectra measurements of surfaces using coherence scanning interferometry *Proc. SPIE* **9890** 98900Q

Colonna de Lega X 2004 Aberration characterization using frequency domain analysis of low-coherence interferograms *Proc. SPIE* **5531** 208

Colonna de Lega X and de Groot P 2008 Characterization of materials and film stacks for accurate surface topography measurement using a white-light optical profiler *Proc. SPIE* **6995** 69950P

Colonna de Lega X and de Groot P 2012 Lateral resolution and instrument transfer function as criteria for selecting surface metrology instruments *Imaging and Applied Optics Technical Digest* OTu1D.4

Coupland J M and Lobera J 2008 Holography, tomography and 3D microscopy as linear filtering operations *Meas. Sci. Technol.* **19** 074012

Coupland J M and Lobera J 2010 Measurement of steep surfaces using white light interferometry *Strain* **46** 69–78

Coupland J M, Mandal R, Palodhi K and Leach R K 2013 Coherence scanning interferometry: linear theory of surface measurement *Appl. Opt.* **52** 3662–70

Coupland J M and Nikolaev N I 2019 A new approach to vector scattering: the 3s boundary source method *Opt. Express* **27** 30380

Damian V S, Bojan M, Schiopu P, Iordache I, Ionita B and Apostol D 2009 White light interferometry applications in nanometrology *Proc. SPIE* **7297** 72971H

de Groot P 2006 Stroboscopic white-light interference microscopy *Appl. Opt.* **45** 5840–4

de Groot P 2011 Coherence scanning interferometry *Optical Measurement of Surface Topography* ed R K Leach (Berlin: Springer) ch 9

de Groot P 2015a Principles of interference microscopy for the measurement of surface topography *Adv. Opt. Photonics* **7** 1

de Groot P 2015b Interference microscopy for surface structure analysis *Handbook of Optical Metrology: Principles and Applications* 2nd edn (Boca Raton, FL: CRC Press) pp 791–828

de Groot P 2017 The meaning, measure of vertical resolution in optical surface topography measurement *Appl. Sci.* **7** 54

de Groot P J 2019 A review of selected topics in interferometric optical metrology *Rep. Prog. Phys.* **82** 056101

de Groot P and Beverage J 2015 Calibration of the amplification coefficient in interference microscopy by means of a wavelength standard *Proc. SPIE* **9526** 952610

de Groot P and Biegen J F 2016 Interference microscope objectives for wide-field areal surface topography measurements *Opt. Eng.* **55** 074110

de Groot P and Colonna de Lega X 2004 Signal modeling for low-coherence height-scanning interference microscopy *Appl. Opt.* **43** 4821–30

de Groot P and Colonna de Lega X 2007 Angle-resolved three-dimensional analysis of surface films by coherence scanning interferometry *Opt. Lett.* **32** 1638–40

de Groot P and Colonna de Lega X 2008 Transparent film profiling and analysis by interference microscopy *Proc. SPIE* **7064** 70640I

de Groot P and Colonna de Lega X 2020 Fourier optics modeling of interference microscopes *J. Opt. Soc. Am.* A **37** B1

de Groot P, Colonna de Lega X, Kramer J and Turzhitsky M 2002 Determination of fringe order in white-light interference microscopy *Appl. Opt.* **41** 4571

de Groot P, Colonna de Lega X, Liesener J and Darwin M 2008 Metrology of optically-unresolved features using interferometric surface profiling and RCWA modeling *Opt. Express* **16** 3970

de Groot P, Colonna de Lega X, Su R and Leach R K 2019 Does interferometry work? A critical look at the foundations of interferometric surface topography measurement *Proc. SPIE* **11102** 111020G

de Groot P and Deck L 1995 Surface profiling by analysis of white-light interferograms in the spatial frequency domain *J. Mod. Opt.* **42** 389–401

de Groot P and DiSciacca J 2018 Surface-height measurement noise in interference microscopy *Proc. SPIE* **10749** 107490Q

Doi T, Toyoda K and Tanimura Y 1997 Effects of phase changes on reflection and their wavelength dependence in optical profilometry *Appl. Opt.* **36** 7157

Dunsby C, Gu Y and French P M W 2003 Single-shot phase-stepped wide-field coherence-gated imaging *Opt. Express* **11** 105

Eifler M, Ströer F, Hering J, von Freymann G and Seewig J 2019 User-oriented evaluation of the metrological characteristics of areal surface topography measuring instruments *Proc. SPIE* **11056** 110560Y
Ekberg P, Su R and Leach R K 2017 High-precision lateral distortion measurement and correction in coherence scanning interferometry using an arbitrary surface *Opt. Express* **25** 18703
Fay M, Badami V G and Colonna de Lega X 2014c Characterizing additive manufacturing parts using coherence scanning interferometry *Proc. ASPE Spring Topical Meeting* pp 135–8
Fay M, Colonna de Lega X and de Groot P 2014a Measuring high-slope and super-smooth optics with high-dynamic-range coherence scanning interferometry *Classical Optics* OW1B.3
Fay M, Colonna de Lega X and Schmidt M 2014b Measuring high-slope parts using coherence scanning interferometry *Proc. ASPE Annual Meeting* pp 604–8
Fay M and Dresel T 2017 Applications of model-based transparent surface films analysis using coherence-scanning interferometry *Opt. Eng.* **56** 111709
Feng X, Senin N, Su R, Ramasamy S and Leach R K 2019 Optical measurement of surface topographies with transparent coatings *Opt. Lasers Eng.* **121** 261–70
Ferreras Paz V, Peterhänsel S, Frenner K and Osten W 2012 Solving the inverse grating problem by white light interference Fourier scatterometry *Light: Sci. Appl.* **1** 1–7
Freischlad K 2012 Sub-angstrom surface metrology with a virtual reference interferometer *Proc. SPIE* **8493** 84930B
Ghim Y S and Davies A 2012 Complete fringe order determination in scanning white-light interferometry using a Fourier-based technique *Appl. Opt.* **51** 1922–8
Gianto G, Salzenstein F and Montgomery P 2016 Comparison of envelope detection techniques in coherence scanning interferometry *Appl. Opt.* **55** 6763
Giusca C L and Leach R K 2013 Calibration of the scales of areal surface topography measuring instruments: part 3. Resolution *Meas. Sci. Technol.* **24** 105010
Giusca C L, Leach R K, Helary F, Gutauskas T and Nimishakavi L 2012a Calibration of the scales of areal surface topography-measuring instruments: part 1. Measurement noise and residual flatness *Meas. Sci. Technol.* **23** 035008
Giusca C L, Leach R K and Helery F 2012b Calibration of the scales of areal surface topography measuring instruments: part 2. Amplification, linearity and squareness *Meas. Sci. Technol.* **23** 065005
Gomez C, Campanelli C, Su R and Leach R K 2020a Surface-process correlation for an ink-jet printed transparent fluoroplastic *Surf. Topogr. Metrol. Prop.* **8** 034002
Gomez C, Su R, de Groot P and Leach R K 2020b Noise reduction in coherence scanning interferometry for surface topography measurement *Nanomanuf. Metrol.* **3** 68–76
Gomez C, Su R, Thompson A, DiSciacca J, Lawes S and Leach R K 2017 Optimization of surface measurement for metal additive manufacturing using coherence scanning interferometry *Opt. Eng.* **56** 111714
Goodman J W 2017 *Introduction to Fourier Optics* 4th edn (New York: Freeman)
Guo T, Ma L, Chen J, Fu X and Hu X 2011 Microelectromechanical systems surface characterization based on white light phase shifting interferometry *Opt. Eng.* **50** 053606
Gu M 2000 *Advanced Optical Imaging Theory* vol 75 (Berlin: Springer)
Hamprecht F A 2005 Bayesian surface estimation for white light interferometry *Opt. Eng.* **44** 015601
Hanhijärvi K, Kassamakov I, Aaltonen J, Heikkinen V, Sainiemi L, Franssila S and Hæggström E 2013 Through-silicon stroboscopic characterization of an oscillating MEMS thermal actuator using supercontinuum interferometry *IEEE/ASME Trans. Mechatron.* **18** 1418–20

Harasaki A, Schmit J and Wyant J C 2000 Improved vertical-scanning interferometry *Appl. Opt.* **39** 2107
Harasaki A, Schmit J and Wyant J C 2001 Offset of coherent envelope position due to phase change on reflection *Appl. Opt.* **40** 2102
Harasaki A and Wyant J C 2000 Fringe modulation skewing effect in white-light vertical scanning interferometry *Appl. Opt.* **39** 2101
Hariharan P 2007 White-light interference microscopy *Basics of Interferometry* (Amsterdam: Elsevier) pp 105–10
Haitjema H and Morel M A A 2005 Noise bias removal in profile measurements *Measurement* **38** 21–9
Heikkinen V, Kassamakov I, Paulin T, Nolvi A and Hæggström E 2013 Stroboscopic scanning white light interferometry at 27 MHz with 16 μm coherence length using a non-phosphor LED source *Opt. Express* **21** 5247
Henning A, Giusca C, Forbes A, Smith I, Leach R K, Coupland J and Mandal R 2013 Correction for lateral distortion in coherence scanning interferometry *Ann. CIRP* **62** 547–50
Hovis C, Shahinian H and Evans C 2019 Observations on the effect of retrace error in scanning white light interferometry of smooth optical surfaces *Proc. Optical Design and Fabrication* **F161** OM4A.2
Huang Y, Gao J, Zhang L, Deng H and Chen X 2020 Fast template matching method in white-light scanning interferometry for 3D micro-profile measurement *Appl. Opt.* **59** 1082
Jeon J W, Jeong H W, Jeong H B and Joo K-N 2019 High-speed polarized low coherence scanning interferometry based on spatial phase shifting *Appl. Opt.* **58** 5360
Joo W-D, Kim S, Park J, Lee K, Lee J, Kim S, Kim Y-J and Kim S-W 2013 Femtosecond laser pulses for fast 3-D surface profilometry of microelectronic step-structures *Opt. Express* **21** 15323
Kassamakov I, Hanhijärvi K, Abbadi I, Aaltonen J, Ludvigsen H and Hæggström E 2009 Scanning white-light interferometry with a supercontinuum source *Opt. Lett.* **34** 1582
Kino G S and Chim S S C 1990 Mirau correlation microscope *Appl. Opt.* **29** 3775
Kovesi P 2015 Good colour maps: how to design them, *arXiv*: 1509.03700 [cs.GR]
Laopornpichayanuwat W, Visessamit J and Tianprateep M 2012 3-D surface roughness profile of 316-stainless steel using vertical scanning interferometry with a superluminescent diode *Measurement* **45** 2400–6
Larkin K G 1996 Efficient nonlinear algorithm for envelope detection in white light interferometry *J. Opt. Soc. Am.* A **13** 832
Lazzini G, Romoli L, Blunt L and Gemini L 2017 Design and characterization of textured surfaces for applications in the food industry *Surf. Topogr.* **5** 044005
Leach R K, Bourell D, Carmignato S, Donmez A, Senin N and Dewulf W 2019 Geometrical metrology for metal additive manufacturing *Ann. CIRP* **68** 677–700
Leach R K, Giusca C L, Haitjema H, Evans C and Jiang X 2015 Calibration and verification of areal surface texture measuring instruments *Ann. CIRP* **64** 797–813
Leach R K, Sims-Waterhouse D, Medeossi F, Savio E, Carmignato S and Su R 2018 Fusion of photogrammetry and coherence scanning interferometry data for all-optical coordinate measurement *Ann. CIRP* **67** 599–602
Lehmann P 2010 Vertical scanning white-light interference microscopy on curved microstructures *Opt. Lett.* **35** 1768–70
Lehmann P, Kühnhold P and Xie W 2014 Reduction of chromatic aberration influences in vertical scanning white-light interferometry *Meas. Sci. Technol.* **25** 065203

Lehmann P, Tereschenko S, Allendorf B, Hagemeier S and Hüser L 2019 Spectral composition of low-coherence interferograms at high numerical apertures *J. Eur. Opt. Soc. Publ.* **15** 5

Lehmann P, Xie W, Allendorf B and Tereschenko S 2018 Coherence scanning and phase imaging optical interference microscopy at the lateral resolution limit *Opt. Express* **26** 7376

Liu M, Cheung C F, Ren M and Cheng C-H 2015 Estimation of measurement uncertainty caused by surface gradient for a white light interferometer *Appl. Opt.* **54** 8670

Lu Y, Park J, Yu L and Kim S-W 2018 3D profiling of rough silicon carbide surfaces by coherence scanning interferometry using a femtosecond laser *Appl. Opt.* **57** 2584

Ma S, Quan C, Zhu R, Tay C J, Chen L and Gao Z 2011 Application of least-square estimation in white-light scanning interferometry *Opt. Lasers Eng.* **49** 1012–8

Mabwa D, Kubiena T, Parnell H, Su R, Furniss D, Tang Z, Leach R K, Benson T M, Scotchford C A and Seddon A B 2020 Evaluating the cytotoxicity of Ge–Sb–Se chalcogenide glass optical fibres on 3T3 mouse fibroblasts *RSC Adv.* under review

Mandal R, Coupland J, Leach R K and Mansfield D 2014 Coherence scanning interferometry: measurement and correction of three-dimensional transfer and point-spread characteristics *Appl. Opt.* **53** 1554

Maniscalco B, Kaminski P M and Walls J M 2014 Thin film thickness measurements using scanning white light interferometry *Thin Solid Films* **550** 10–6

Mansfield D 2008 Extraction of film interface surfaces from scanning white light interferometry *Proc. SPIE* **7101** 71010U

McCutchen C W 1964 Generalized aperture and the three-dimensional diffraction image *J. Opt. Soc. Am.* **54** 240

Moharam M G and Gaylord T K 1983 Three-dimensional vector coupled-wave analysis of planar-grating diffraction *J. Opt. Soc. Am.* **73** 1105

Montgomery P C, Salzenstein F, Montaner D, Serio B and Pfeiffer P 2013 Implementation of a fringe visibility based algorithm in coherence scanning interferometry for surface roughness measurement *Proc. SPIE* **8788** 87883G

Mukhtar H, Montgomery P, Gianto and Susanto K 2016 Rock surface roughness measurement using CSI technique and analysis of surface characterization by qualitative and quantitative results *IOP Conf. Ser.: Earth Environ. Sci.* **29** 012028

Mun J I, Jo T, Kim T and Pahk H J 2015 Residual vibration reduction of white-light scanning interferometry by input shaping *Opt. Express* **23** 464

Nakano K, Hane K, Okuma S and Eguchi T 1997 Visualization of 50 MHz surface acoustic wave propagation using stroboscopic phase-shift interferometry *Opt. Rev.* **4** 265–9

Niehues J, Lehmann P and Bobey K 2007 Dual-wavelength vertical scanning low-coherence interferometric microscope *Appl. Opt.* **46** 7141–8

Oh J S and Kim S-W 2005 Femtosecond laser pulses for surface-profile metrology *Opt. Lett.* **30** 2650

Olszak A 2000 Lateral scanning white-light interferometer *Appl. Opt.* **39** 3906

Palodhi K, Coupland J and Leach R K 2011 Absolute surface topography measurement of composite structures using coherence scanning interferometry *Proc. 11th EUSPEN* pp 255–8

Pavliček P and Hýbl O 2012 White-light interferometry on rough surfaces-measurement uncertainty caused by noise *Appl. Opt.* **51** 465–73

Pavliček P and Kučera J 2019 Coherence scanning interferometry with a focus-tunable lens *Appl. Opt.* **58** G91

Pavliček P and Michálek V 2012 White-light interferometry—envelope detection by Hilbert transform and influence of noise *Opt. Lasers Eng.* **50** 1063–8

Pavliček P and Mikeska E 2020 White-light interferometer without mechanical scanning *Opt. Lasers Eng.* **124** 105800

Pecheva E, Montgomery P, Montaner D and Pramatarova L 2007 White light scanning interferometry adapted for large-area optical analysis of thick and rough hydroxyapatite layers *Langmuir* **23** 3912–8

Pförtner A and Schwider J 2001 Dispersion error in white-light Linnik interferometers and its implications for evaluation procedures *Appl. Opt.* **40** 6223

Raid I, Eifler M, Kusnezowa T and Seewig J 2015 Calibration of ellipso-height-topometry with nanoscale gratings of varying materials *Optik* **126** 4591–6

Roy M, Cooper I, Moore P, Sheppard C J R and Hariharan P 2005 White-light interference microscopy: effects of multiple reflections within a surface film *Opt. Express* **13** 164

Roy M, Schmit J and Hariharan P 2009 White-light interference microscopy: minimization of spurious diffraction effects by geometric phase-shifting *Opt. Express* **17** 4495–9

Roy M, Svahn P, Cherel L and Sheppard C J R 2002 Geometric phase-shifting for low-coherence interference microscopy *Opt. Lasers Eng.* **37** 631–41

Sachs R and Stanzel F 2014 Interference microscopy for clean air—how optical metrology is improving quality control of fuel injection systems *Proc. Fringe* **2013** 535–8

Sandoz P 1997 Wavelet transform as a processing tool in white-light interferometry *Opt. Lett.* **22** 1065

Schmit J, Creath K and Wyant J C 2007 Surface profilers, multiple wavelength, and white light intereferometry *Optical Shop Testing* 3rd edn ed D Malacara (Hoboken, NJ: Wiley) pp 667–755

Senin N, Thompson A and Leach R K 2017 Characterisation of the topography of metal additive surface features with different measurement technologies *Meas. Sci. Technol.* **28** 095003

Sergeeva M, Khrenikov K, Hellmuth T and Boerret R 2010 Sub surface damage measurements based on short coherent interferometry *J. Eur. Opt. Soc. Rapid Publ.* **5** 10003

Sheppard C J R and Larkin K G 1995 Effect of numerical aperture on interference fringe spacing *Appl. Opt.* **34** 4731

Simonsen I 2010 Optics of surface disordered systems: a random walk through rough surface scattering phenomena *Eur. Phys. J.: Spec. Top.* **181** 1–103

Song Z, Guo T, Fu X and Hu X 2018 Residual vibration control based on a global search method in a high-speed white light scanning interferometer *Appl. Opt.* **57** 3415

Speidel A, Su R, Mitchell-Smith J, Dryburgh P, Bisterov I, Pieris D, Li W, Patel R, Clark M and Clare A T 2018 Crystallographic texture can be rapidly determined by electrochemical surface analytics *Acta Mater.* **159** 89–101

Su R, Coupland J M, Wang Y and Leach R K 2017a Tolerance on sphere radius for the calibration of the transfer function of coherence scanning interferometry *Proc. SPIE* **10329** 103290L

Su R, Coupland J M, Sheppard C J R and Leach R K 2020a Scattering and three-dimensional imaging in surface topography measuring interference microscopy *J. Opt. Soc. Am. A* Under review

Su R, Thomas M, Leach R K and Coupland J 2018 Effects of defocus on the transfer function of coherence scanning interferometry *Opt. Lett.* **43** 82

Su R, Thomas M, Liu M, Drs J, Bellouard Y, Pruss C, Coupland J and Leach R K 2020b Lens aberration compensation in interference microscopy *Opt. Lasers Eng.* **128** 106015

Su R, Wang Y, Coupland J and Leach R K 2017b On tilt and curvature dependent errors and the calibration of coherence scanning interferometry *Opt. Express* **25** 3297

Thomas M, Su R, de Groot P and Leach R K 2020a Optical topography measurement of steeply-sloped surfaces beyond the specular numerical aperture limit *Proc. SPIE* **11352** 1135207

Thomas M, Su R, Nikolaev N, Coupland J and Leach R K 2020b Modeling of interference microscopy beyond the linear regime *Opt. Eng.* **59** 1

Townsend A, Senin N, Blunt L, Leach R K and Taylor J S 2016 Surface texture metrology for metal additive manufacturing: a review *Precis. Eng.* **46** 34–47

Ullmann V, Emam S and Manske E 2015 White-light interferometers with polarizing optics for length measurements with an applicable zero-point detection *Meas. Sci. Technol.* **26** 84010

Vo Q, Fang F, Zhang X and Gao H 2017 Surface recovery algorithm in white light interferometry based on combined white light phase shifting and fast Fourier transform algorithms *Appl. Opt.* **56** 8174

Wang J, Su R, Leach R K, Lu W, Zhou L and Jiang X 2018 Resolution enhancement for topography measurement of high-dynamic-range surfaces via image fusion *Opt. Express* **26** 34805

Watanabe K, Ohshima M and Nomura T 2014 Simultaneous measurement of refractive index and thickness distributions using low-coherence digital holography and vertical scanning *J. Opt.* **16** 045403

Wiersma J T and Wyant J C 2013 Vibration insensitive extended range interference microscopy *Appl. Opt.* **52** 5957–61

Wiesner B, Hybl O and Häusler G 2012 Improved white-light interferometry on rough surfaces by statistically independent speckle patterns *Appl. Opt.* **51** 751–7

Xie W, Lehmann P and Niehues J 2012 Lateral resolution and transfer characteristics of vertical scanning white-light interferometers *Appl. Opt.* **51** 1795

Xie W, Lehmann P, Niehues J and Tereschenko S 2016 Signal modeling in low coherence interference microscopy on example of rectangular grating *Opt. Express* **24** 14283

Yoshino H, Abbas A, Kaminski P M, Smith R, Walls J M and Mansfield D 2017 Measurement of thin film interfacial surface roughness by coherence scanning interferometry *J. Appl. Phys.* **121** 105303

Yoshino H, Kaminski P M, Smith R, Walls J M and Mansfield D 2016 Refractive index determination by coherence scanning interferometry *Appl. Opt.* **55** 4253

Yu Y and Mansfield D 2015 Characterisation of thin films using a coherence scanning interferometry *J. Mater. Sci. Chem. Eng.* **03** 15–21

Zou Y, Li Y, Kaestner M and Reithmeier E 2016 Low-coherence interferometry based roughness measurement on turbine blade surfaces using wavelet analysis *Opt. Lasers Eng.* **82** 113–21

Chapter 3

Focus variation

Claudia Repitsch, Kerstin Zangl, Franz Helmli and Reinhard Danzl

Focus variation enables the measurement of texture, form and geometry, also allowing for the measurement of complex geometries of micro-scale samples with nanometre vertical discrimination. In this chapter, advances in focus variation-based measurement principles are presented, with improvements being achieved in three areas of application: (i) modern focus variation instruments can now measure optically smooth surfaces; (ii) 'vertical focus probing' is a newly developed extension to focus variation and allows the direct measurement of vertical walls and holes; and (iii) focus variation instruments that are suitable for in-process surface measurements. First, the basic principles of focus variation are briefly summarised, followed by a description of these new improvements and a case study for each.

3.1 Introduction to focus variation

Focus variation (FV) is an optical measurement technology for surface topography measurement. FV is suitable for high-resolution areal surface texture measurement, coordinate metrology and the measurement of high slope angles. The areas of application range from the tool industry, micro-precision manufacturing and additive manufacturing, to the automotive, aerospace and medical industries.

Figure 3.1 shows some examples of 3D measurements obtained with FV.

FV uses the small depth of focus of an optical instrument and combines it with vertical scanning to provide topographic and colour information from the variation of focus (Danzl *et al* 2011, Leach 2011, Leach 2014, Nikolaev *et al* 2016, Piska and Metelkova 2014, Schuth and Buerakov 2017, Tian *et al* 2013, ISO 2015). The principle was first published in 1924 (von Helmholtz 1924), however, its use in high-precision dimensional metrology is relatively new compared to other optical measurement methods. An extensive overview of FV, including good practice guidance, can be found in Helmli (2011) and it is assumed the reader is familiar with this publication, although some basic principles are given here. Following Helmli's publication in 2011, an ISO specification standard on FV instruments was

doi:10.1088/978-0-7503-2528-8ch3

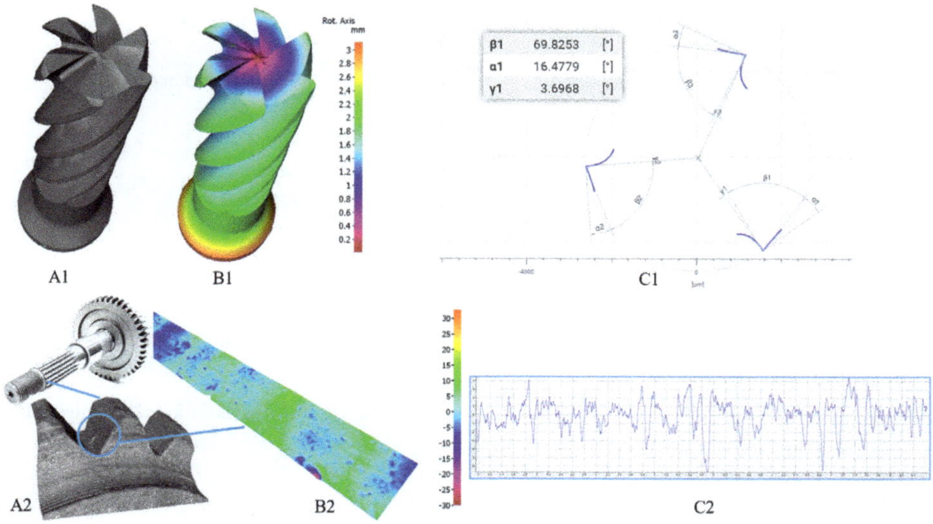

Figure 3.1. 3D measurements obtained with FV. Top: (A1) Measurement of a milling cutter, (B1) 3D dataset in pseudo-colours, and (C1) measurement of clearance angle (α), wedge angle (β) and rake angle (γ). Bottom: (A2) 3D measurement on a gear shaft, (B2) one tooth flank in pseudo-colours and (C2) roughness profile on a tooth flank.

published in 2015 (ISO 2015). The ISO standard describes the metrological characteristics as well as the principle of the FV technology. In this section, a short overview of the FV principle is given. In section 3.2, the following three new approaches based on FV are described:

1. The use of an extension to FV for the measurement of smooth surfaces.
2. The use of an extension to FV for the measurement of vertical walls and holes (so-called 'vertical focus probing').
3. The use of FV for in-process measurements.

3.1.1 Basic principles of focus variation

A schematic diagram of a typical measurement instrument based on FV is shown in figure 3.2. The main components are:

- an optical instrument with various lenses and other optical components;
- the illumination source(s), for example, coaxial light or ring light;
- a digital sensor/array detector; and
- a scanning unit to change the distance between the object and the optical set-up.

Measurement instruments based on FV use the following measurement process. Light from a broadband, white-light source (or alternatively a red, green or blue LED if no colour information is required) is inserted into the optical path of the instrument and projected onto the sample. The light is incident on the surface of the sample and is scattered. Depending on the surface topography, diffuse and specular

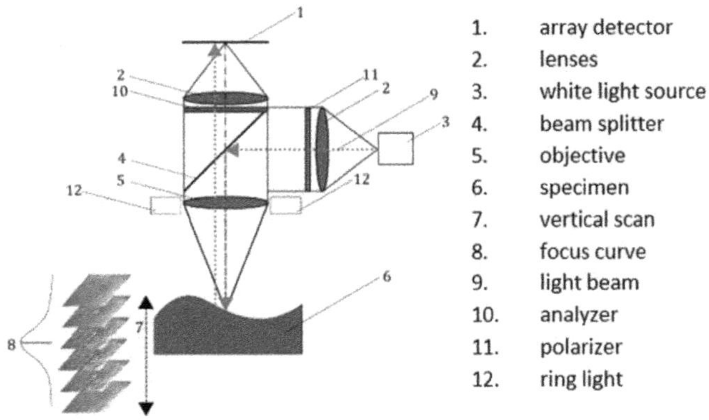

Figure 3.2. Schematic diagram of a typical measurement device based on focus variation. Reprinted/adapted by permission from Springer Nature: Helmli 2011.

reflection occur in variable degrees. A combination of both types of reflection is common in general (see Nikolaev *et al* (2016) for an in-depth treatment of the theory of FV). The reflected light is collected by the objective lens and is projected to a digital sensor/array detector. Depending on the vertical distance of the sample to the objective lens, the light is more or less focused on the digital sensor. Due to the small depth of field of the optics, only small regions of the sample are highly focused on the sensor.

A complete areal measurement of the surface is achieved by performing a vertical scan. The optics or the sample are moved vertically along the optical axis while continuously capturing data from the digital sensor. Due to the vertical motion, different regions of the sample are focused. This results in a change in contrast in the images captured on the digital sensor. By analysing the data for each point in the captured image, a focus information curve along the vertical axis is generated. Algorithms convert the focus information curves (one for each sampling point) into a 3D dataset and a true colour image with full depth of field. The true colour image can help with the measurement and identification of distinctive local surface features. The visual correlation between the true colour image of the sample surface and its depth information are often linked to each other and can be an essential aspect of 3D measurement (Leach 2011). As can be seen from figure 3.2, unlike other optical measurement principles, such as confocal (see chapter 4) or coherence scanning interferometry (see chapter 2), FV is not limited to coaxial illumination, which allows the use of many different light sources, for example, coaxial, ring light, polarisation and even transmitted light. Therefore, FV has been shown to allow measurement of slope angles up to 87° (Danzl *et al* 2011).

3.1.2 Algorithms for focus variation

During the vertical scan of a sample, data/images are continuously captured on the digital sensor. In general, if a sample point is out of focus, the point and its neighbouring points will produce almost identical grey values on the digital sensor (assuming no other light effects this scenario). If the sample point is sharply focused, then the neighbouring points will have a larger variation of grey values (assuming

Scan position	Surface image	Standard deviation
Out of focus	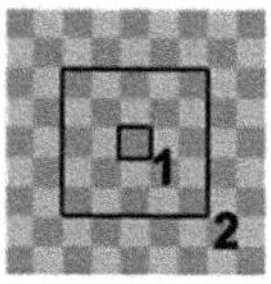	10
Almost in focus		20
In focus		50
Almost in focus		20
Out of focus	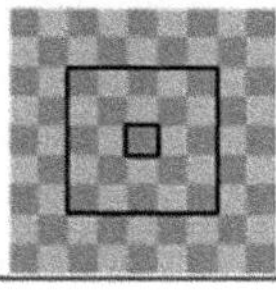	10

Figure 3.3. Sample grey values and their standard deviation at different vertical scan positions. Region 1: point of interest for which the focus information is calculated. Region 2: 5 × 5 neighbourhood of points around the point of interest. Reprinted/adapted by permission from Springer Nature: Helmli 2011.

the object is not homogeneous or completely smooth). Hence, in order to measure the contrast (or focus information) in a captured image, a small region around the current pixel position is considered. One simple approach is the use of the standard deviation of the grey values in a small neighbourhood of the point of interest. This is demonstrated in figure 3.3 for a chess-pattern-like sample at five different scan positions. At each scan position, the standard deviation of the grey values in a 5 × 5 pixel neighbourhood of the sample point is determined. The focus information curve is built from these standard deviation values, and by calculating the maximum of the curve, the height value of the sample point can be determined. In the chess pattern example, at the start of vertical scanning, the considered sample point (point of interest) is out of focus, then it comes into focus and then out of focus again. This behaviour is numerically represented by the standard deviation values (see figure 3.3). A wide variety of algorithms and operators have been proposed to measure the degree of focus of either a whole image or a pixel position. In Pertuz *et al* (2013) a good overview is given; different focus measures (grouped into six categories) are presented and their performance is assessed.

Table 3.1. Methods for calculating the maximum of the focus information curve.

Method	Accuracy	Speed
Polynomial	High	Medium
Curve fitting	High	Slow
Peak	Low	Fast

In table 3.1, several methods for calculating the maximum of the focus curve are listed together with their speed and accuracy. The fastest, but least accurate method is to simply use the scan position with the maximum. Using this method, in the chess pattern example (see figure 3.3), the scan position with a standard deviation of 50 would be selected. More advanced methods fit functions (polynomials or more complex functions) to the (discrete) focus information curve and use the maximum of the fitted function to determine the corresponding height value/z-position.

3.2 Advances in focus variation

The measurement of highly smooth surfaces and the measurement of micro-holes and vertical walls with slope angles beyond 87° are challenging tasks that are relevant for many applications, for example, the inspection of micro-injection holes of valves, bidirectional dimensional measurements by lateral probing according to ISO 10360 part 8 (ISO 2013) or the measurement of highly polished medical implants and cutting tools.

Moreover, with the ongoing trend in smart manufacturing, the need to use a measurement instrument not only in the laboratory environment but to integrate it directly in the manufacturing process is inevitable and critical for many applications. For such in-process measurements, requirements for measurement instruments and measurement technologies are challenging. On the one hand, environmental conditions must be handled robustly, for example, temperature variations and vibration. On the other hand, flexible automatic measurement solutions must be developed (see chapter 7 for an in-depth discussion of the challenges and implementations of in-process optical measurement, including FV).

In order to address these challenges, FV technology and measurement instruments have been further developed in recent years. This section and the associated case studies in section 3.3 discuss the three main advances in FV in detail—the measurement of smooth surfaces, the measurement of steep slopes by lateral probing and in-process measurement.

3.2.1 Measurement of smooth surfaces

FV uses contrast to measure depth. Contrast is a result of the image formation process of the surface texture including the surface roughness (the high-spatial frequency components) of the sample. If there is no image contrast, because the sample is very smooth, then FV in general will not give useful results. In practice, a 'golden rule' has been established by many manufacturers: surfaces must have at least an $Ra \geqslant 9$ nm with $\lambda c = 2\ \mu$m (see Leach 2014, Danzl *et al* 2011 for parameter and filtering descriptions) for high-resolution surface texture measurements using a FV measurement instrument. However, several developments to overcome this limitation will be discussed.

In coordinate metrology, it is common to use multiple sensors in one measuring instrument to select the most appropriate probing system for each application. In optical metrology, developments move in the same direction (see Matilla *et al* 2016, Wang *et al* 2015). To measure complex geometries and smooth surfaces at the same time, modern optical measurement instruments combine two or more optical measurement technologies, such as confocal microscopy or structured illumination and focus variation into one device (see Schuth and Buerakov 2017 and chapter 4). Thus, depending on the application, a different measurement principle is applied; for the measurement of steep slopes, FV is used, for the evaluation of surface roughness, a confocal measurement is performed. The registration of these measurements into one coordinate system allows the evaluation in one dataset.

One recently introduced approach for measuring smooth surfaces with FV is to employ modulated illumination during the vertical scanning process, where the intensity modulation is modelled as a function of time and lateral position at the same time. Thus, each individual measurement point is optimally illuminated in order to obtain robust and high-resolution 3D depth data. While a single measurement point is illuminated with varying intensity at different points in time, two measurement points are illuminated with varying illumination at the same time. The result is a measurement that is more robust and has higher depth resolution, but also a significantly more robust and higher lateral resolution, plus measurement noise can be reduced (see section 3.3.1.1).

A measurement of an optically smooth mirror using FV is shown in figure 3.4. The profile and surface roughness parameter values are in the range of a few nanometres (Ra = 2.7 nm and Sa = 3.6 nm). Such a measurement would not be possible with conventional FV.

Active illumination FV was proposed by Bermudez *et al* (2019). The principle was introduced by Noguchi and Nayar (1994) and has been further developed for use in a confocal microscope. Active illumination FV allows the measurement of smooth surfaces by projecting a pattern onto the surface to generate texture on the surface and consequently contrast in the image (see chapter 4).

3.2.2 Measurement of vertical walls and holes using vertical focus probing

The measurement of components with complex geometries often includes the measurement of steep slopes and inner geometries such as micro-holes, where topography is essential for their functionality. Well-known industrial applications include the measurement of micro-scale injection holes of valves in the automotive industry (see section 3.3.2.2 for a case study) or the measurement of micro-cooling holes on turbine blades in the aerospace industry. In dimensional metrology, the verification of geometric dimensions is a common task, for example, the measurement of the lateral distance between two opposite vertical walls. In section 3.1, a brief introduction to the measurement principle of FV has been given, showing

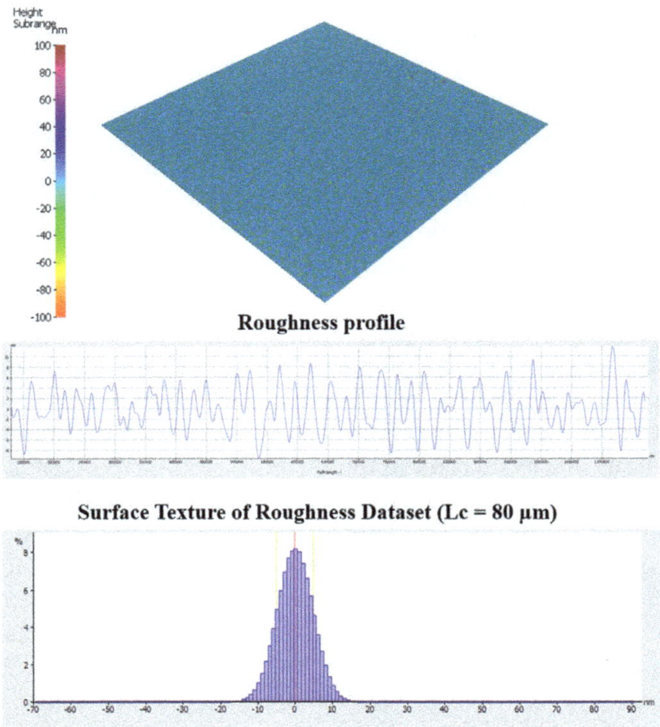

parameter	value [nm]
Ra	2.7
Rq	3.2
Rz	10.3

parameter	value [nm]
Sa	3.4
Sq	4.2
Sz	53.4

Figure 3.4. Measurement of a mirror using modulated illumination FV.

measurement capabilities on slope angles up to 87° (Danzl *et al* 2011). Using standard FV, the measurement of larger slope angles cannot be performed as straightforwardly as with tactile instruments—the sample and/or the sensor has to be articulated to change the viewing angle with respect to the measured surface, for example, by using a rotation and tilt unit or by rotating the sensor. In the centre image of figure 3.5, this approach is illustrated for the measurement of a lateral distance *d*.

In Zangl *et al* (2019), the measurement principle of vertical focus probing (VFP) was introduced. VFP is an extension to FV and an entirely optical measurement technology. VFP allows the measurement of slope angles greater than 87° with optical 3D measurement instruments and, therefore, the measurement of vertical walls by lateral probing. This new approach is illustrated in the right image of figure 3.5. It should be noted that a video coordinate measuring system (CMS) can also measure the width of a part, but for this 2D type of measurement, the illumination strongly affects the width measurement (Coveney 2014).

3.2.2.1 VFP measurement principle

As with FV, VFP utilises the reflective properties of the surface. Depending on the topography of the measured part, the light is diffracted into several directions when it is incident on a surface through the objective. In the case of ideal Lambertian reflectance, the light is reflected at equal intensity into each direction, whereas in the case of specular reflection, the light is scattered in one direction according to the law

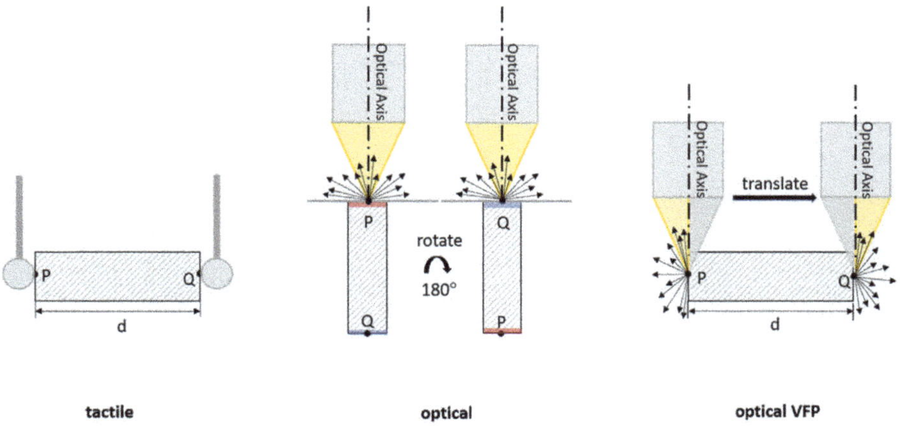

Figure 3.5. Principle of the measurement of a lateral distance *d* for tactile measurement instruments (left), optical measurement instruments based on FV with the use of a rotation system (centre) and optical measurement instruments based on vertical focus probing (right).

of reflection. Typical surfaces have both diffuse and specular reflective properties, which are a result of the spatial frequency distribution of the topography and the incident temporal spectrum. As discussed in section 3.1, measurement instruments based on FV are not restricted to a specific illumination system. Most importantly, depending on the type and position of the illumination, the geometry of the sample and the texture properties of the surface, some part of the light is almost always reflected back to the objective lens, even if the measured surface is parallel to the optical axis, as is the case for a vertical wall or hole. The light rays are bundled in the optics and gathered by a light sensitive sensor. The intensity of light is reduced by vignetting which is caused by the geometry of the sample—this effect is shown in figure 3.6—on the left side for a vertical wall with a slope angle of 90° and on the right side for a wall with a slope angle greater than 90°.

Using FV, the sample is scanned vertically (in the z-direction) and for each xy-position the maximum of the corresponding focus information curve defines its unique z-position. In the case of VFP, the measurement sample is also scanned vertically, however, at each vertical scan position, the captured data are analysed and focused xy-positions, with the associated z-values, are determined. Hence, each xy-position can have many z-positions, and this opens up the possibility of measuring vertical geometries, such as sidewalls or holes.

The VFP algorithm can be summarised as follows.

1. Move in the vertical direction over a pre-defined scanning range and at each vertical position:
   a. capture image data (see figure 3.7 for a live view at one vertical position);
   b. calculate focus information; and
   c. extract 3D points.
2. Delete all 3D points that do not satisfy some threshold criteria.
3. Apply optical aberration correction (for example, curvature of field correction).

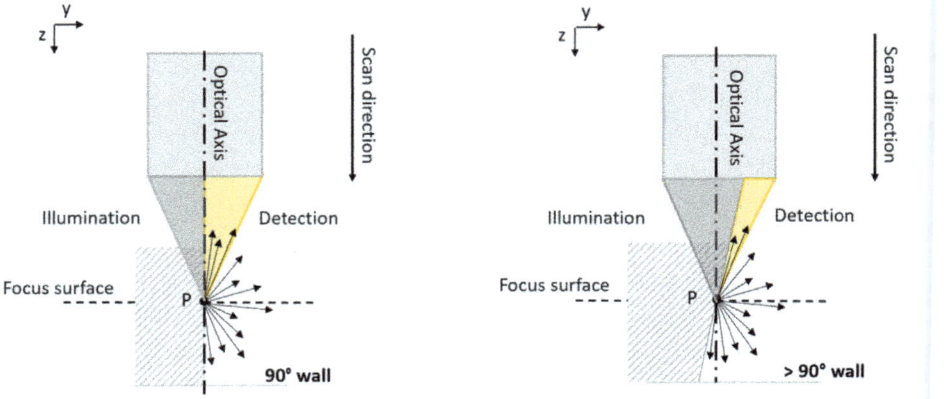

Figure 3.6. Measurement principle of VFP (*yz* view). Left: Partial illumination/detection cone in case of a 90° wall. The orientation of the measured wall is parallel to the optical axis. Right: Partial illumination/detection cone in the case of a wall with more than 90°.

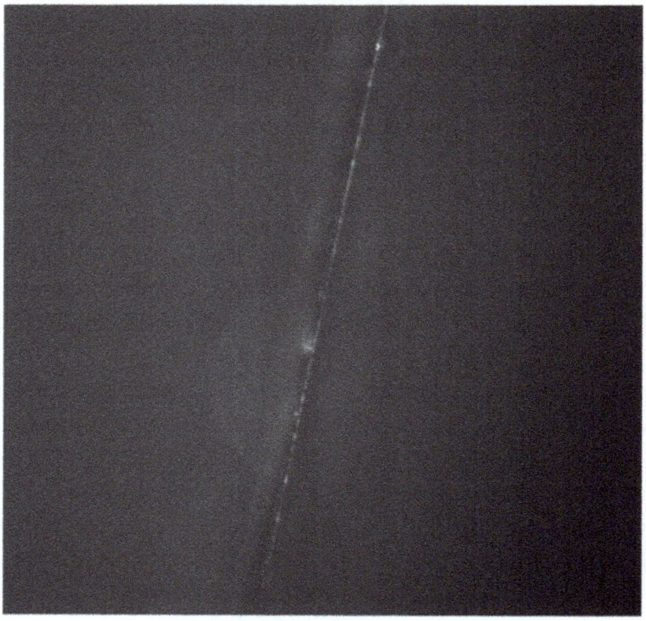

Figure 3.7. Measurement with VFP—live view of a vertical wall at one *z*-position.

The difference in FV assigning exactly one *z*-position to each *xy*-position, and VFP being able to assign arbitrary *z*-positions to each *xy*-position, is clearly visible in the extracted dataset meshes. 3D datasets generated by FV (coherence scanning interferometry and imaging confocal deliver the same type of data) in general

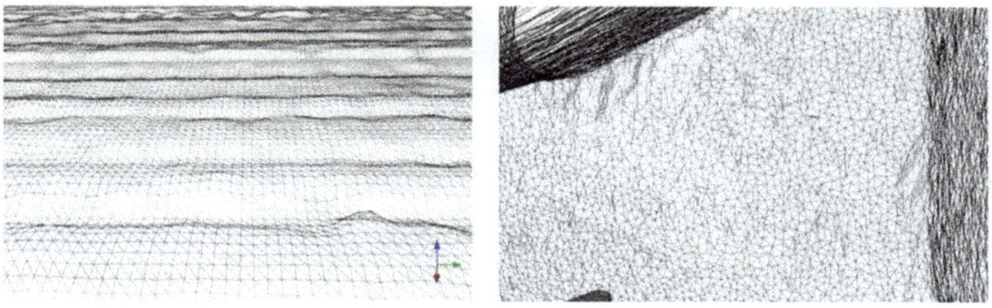

Figure 3.8. Mesh presentation for FV (left image, regular grid) and VFP (right image, arbitrary 3D mesh).

have a regular *xy*-mesh. VFP, however, creates an arbitrary 3D mesh, which consists of a list of 3D points represented by the *x*-, *y*- and *z*-coordinates and a list of triangles. A triangle is typically represented by a triple of indices *i, j* and *k* pointing to the three corners. The distinction between a regular mesh and an arbitrary mesh is illustrated in figure 3.8.

3.2.2.2 Resolution of vertical focus probing

For FV based measurement devices, the optical axis is, in general, approximately perpendicular to the measurement sample. Hence, the illumination/detection cone is not vignetted by the sample (see the left panel of figure 3.9). Let α be the maximum half-angle of the illumination cone and n be the refraction index (assumed equal to unity in the case of air), then the numerical aperture A_N is given by

$$A_N = n \sin \alpha. \tag{3.1}$$

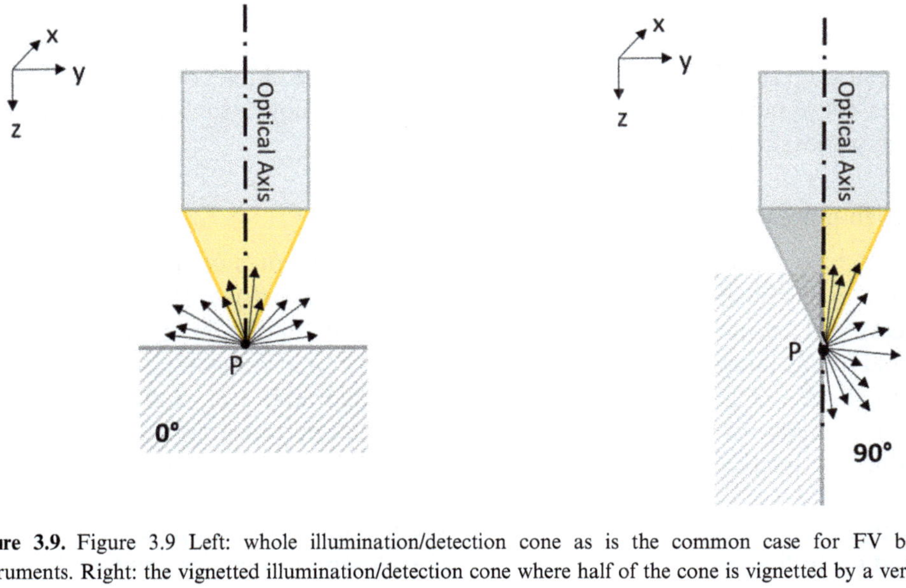

Figure 3.9. Figure 3.9 Left: whole illumination/detection cone as is the common case for FV based instruments. Right: the vignetted illumination/detection cone where half of the cone is vignetted by a vertical wall.

The imaging resolution r and the axial resolution r_z can be described as functions of A_N (see Hecht 2012). The imaging resolution is given by

$$r = \frac{0.61\lambda}{A_N}, \tag{3.2}$$

where λ is the wavelength of the source. The axial resolution is given by

$$r_z = \frac{2\lambda n}{A_N^2}. \tag{3.3}$$

Using VFP, there is a vignetted illumination/detection cone due to the geometry of the measurement part (see the right panel of figure 3.9) and these equations cannot be applied directly. In order to identify the effect of the vignetted illumination/detection cone on the resolution values r and r_z, a numerical point spread function simulation using commercial software was performed for several input parameters.

For comparability, the simulation was performed for two experimental set-ups, once for the whole illumination/detection cone and once for the partial illumination/detection cone, as shown in figure 3.9. Additionally, the resolution values were calculated approximately with the above equations, assuming the full numerical aperture in the x- and z-directions and half the numerical aperture in the y-direction. The results of the numerical simulation show that the resolutions in x and z are not affected by the vignetted illumination/detection cone, but the resolution in the y-direction is reduced by approximately a factor of two. Thus, the simulation results are in good agreement with equation (3.2), when only half the numerical aperture is used for the y-direction in equation (3.2).

It can be concluded that, due to the anisotropic cone, the resolution is different in x and y and reduces in the y-direction, which is most affected by the vignetting. However, the change in resolution has no systematic effect on the position of the calculated 3D point if the optics are optimised in relation to aberrations. Therefore, the reduced resolution in the y-direction has a negligible impact on the measurement results for typical applications in dimensional metrology, such as point-to-point distance measurements.

3.2.3 In-process surface measurements

The ongoing trend to smaller and more accurate structures combined with a high level of precision and efficiency in the production process drives the necessity for in-process surface metrology (see Gao *et al* 2019, and chapter 7). Quality control is carried out as integral part of production, hence the time-consuming step of moving the workpiece to a vibration-free, climate-controlled laboratory for off-line inspection (see figure 3.10) can be impractical. When the workpiece remains in the production line, it is either measured when still mounted on the machine (on-machine, see Gao *et al* 2019 and see figure 3.11) or when it is moved out of the machine (in-line, see Gao *et al* 2019 and see figure 3.12). In-process inspection, including in-line and on-machine, allows for closed-loop control and early defect detection; saving time and costs, as well as improving the production process are

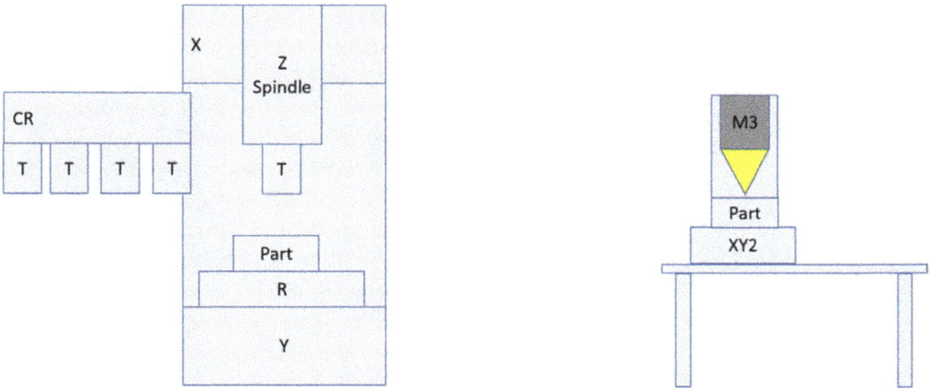

Figure 3.10. Off-line situation where X, Y, Z, R and Spindle are the axes of the production machine, CR is the changer rack, T are the tools for manufacturing, Part is the part of production, M3 is the measurement sensor of the bench-top instrument and XY2 is the *xy* plane.

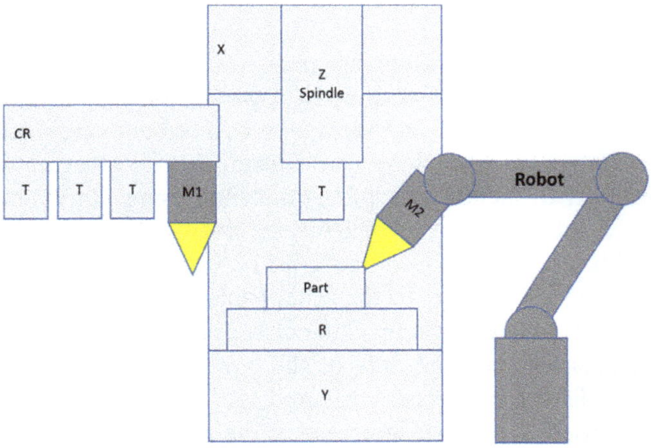

Figure 3.11. In-line and on-machine situation where X, Y, Z, R and Spindle are the axes of the production machine, CR is the changer rack, T are the tools for manufacturing, Part is the part of production, M1 is the measurement sensor in the changer rack (on-machine) and M2 is the measurement sensor on a Robot (in-line).

significant benefits. Traditionally, this in-process metrology was carried out with tactile sensors or optical 2D sensors. Smaller structures and tighter tolerances require optical areal metrology to fulfil the measurement task. Typically, FV, as well as coherence scanning interferometry and confocal techniques, are used in bench-top instruments. These instruments can be placed in the laboratory as well as near a production line. Bringing this optical measurement equipment into the manufacturing process is an important goal in modern manufacturing (Leach 2020).

In the case of FV, off-line measurements, which are measurements that are isolated from the manufacturing environment, are well established. The same is true

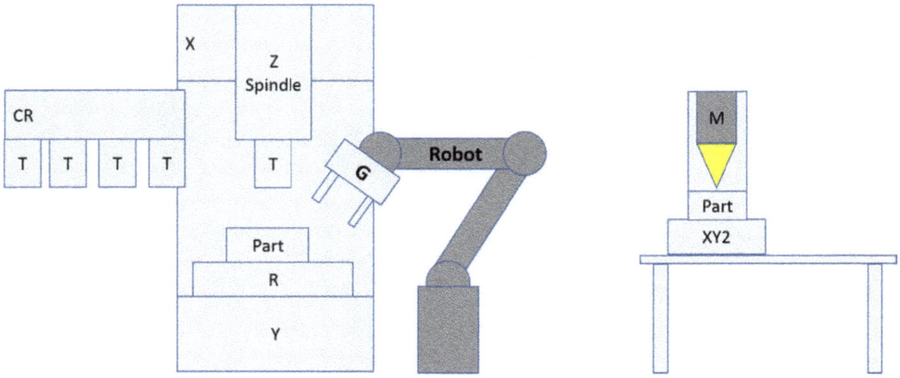

Figure 3.12. In-line and off-machine situation, where X, Y, Z, R and Spindle are the axes of the production machine, CR is the changer rack, T are the tools for manufacturing, Part is the part of production, M is the measurement sensor of the bench-top instrument, XY2 is the *xy*-plane, and G is the gripper on the robot to handle the part between the manufacturing machine and measurement instrument.

for off-machine measurements, which are measurements that are carried out in the manufacturing environment but not directly in the production line. In the following, in-process, including in-line and on-machine, measurements are discussed.

In-line and off-machine metrology (see figure 3.12) involves having a measurement instrument next to the manufacturing machine using, for example, a robot to load the parts from the machine onto the measurement instrument. In the case of in-line and on-machine metrology (see figure 3.11), one option is to have the sensor positioned in a changer rack next to the other tools in the machine. After a specific manufacturing step is finished, the machine changes the tool to the on-machine FV sensor and measures the required area of the part. In such a set-up, both surface texture and small features can be measured. If the axes of the manufacturing machines are also used in the measurement of the part, larger dimensions than the field of view can be measured. In this case, it is important to know the accuracy of the manufacturing axis in order to be able to estimate the uncertainty of the measurement.

In all situations where the FV instrument is used in-line, special requirements must be fulfilled. The requirements for such measurement equipment are high-speed measurement, vibration robustness, robustness in relation to complex surface slopes and materials, and robustness with respect to temperature variations (see chapter 7).

Often, FV instruments are designed for a measurement room with stable temperature (see VDI 2015). A measurement room according to VDI 2617–1 class B keeps the temperature within a range of 0.4 K in one hour, with a gradient of 0.3 K/m. However, in manufacturing environments, higher gradients over time and space are common. In figure 3.13, this is illustrated for a typical working day of a five-axis machine. Eight temperature sensors are positioned in the corners of a cube with 1.5 m side length around an optical 3D CMS, and an additional sensor is placed on the room wall next to the five-axis machine. Between 0900 and 1300, an increase of

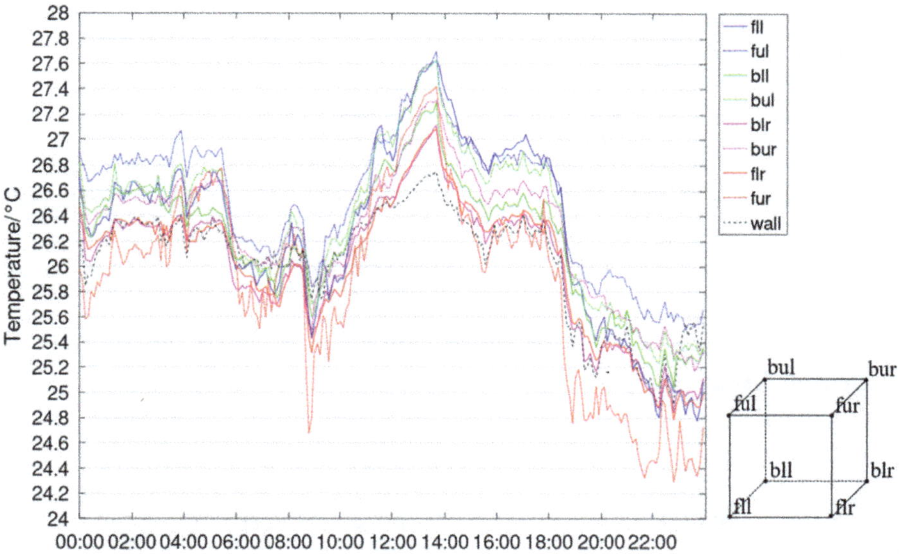

Figure 3.13. Temperature (°C) over time and space in the manufacturing environment of a five-axis machine. The symbols in the box indicate the temperature sensors. The symbol 'wall' denotes the sensor on the wall. The remaining eight symbols denote the sensors in the corners of a cube (side length 1.5 m) around the used optical 3D CMS, as drawn in the sketch of the cube below the symbol box.

1.3° to 2.0° can be seen, and a temperature gradient in space of up to 1.3 K/m can be observed.

Hence, measurement instruments used in manufacturing environments must carefully manage temperature changes. There are three levels of temperature handling:

- not measuring temperature;
- measure temperature and use the information to classify valid and not valid measurements based on the temperature situation during the measurement; and
- measure temperature and use the information to correct the measurement results.

In addition to temperature management, the mechanical design should be invariant to thermal gradients. The use of equal material combinations and of low coefficient of thermal expansion materials, such as Invar, helps to avoid bending and drift effects during the measurement (see chapter 7).

Robustness towards vibration is another important criterion. Confocal, FV and coherence scanning interferometry (CSI) use vertical scanning to obtain 3D data. Since these methods have a different vertical length of influence compared to the measured z-value, they have different sensitivities to vibration. Typically, a longer vertical length of influence leads to lower vibration sensitivity.

FV uses curve fitting over the depth of field. The full width at half maximum (FWHM) of the focus curve is calculated by

$$\mathrm{FWHM_{FV}} = \frac{0.88\lambda}{n - \sqrt{n^2 - A_N{}^2}}. \tag{3.4}$$

For conventional confocal technologies, the FWHM is calculated by (Artigas 2011)

$$\mathrm{FWHM_{confocal}} = \frac{0.64\lambda}{n - \sqrt{n^2 - A_N{}^2}}. \tag{3.5}$$

For CSI, the coherence length l_c determines the vertical length of influence and is calculated by Akcay *et al* (2002)

$$l_c = \frac{4\ln(2)\lambda^2}{\pi\Delta\lambda}, \tag{3.6}$$

where λ is the mean light source wavelength and $\Delta\lambda$ is the spectral width of the light source.

In table 3.2, the vertical length of influence for the three technologies are compared, using the parameter values refractive index $n = 1$, numerical aperture $A_N = 0.3$, mean light source wavelength $\lambda = 500$ nm, and spectral width of the light source in the case CSI is $\Delta\lambda = 150$ nm. The larger the value of the vertical length of influence, the smaller the vibration sensitivity of the respective technology. In general, FV is one of the more robust technologies.

Other important aspects of on-machine metrology are the size of the sensor, the power supply and the enclosure for the sensor to keep it free from various environmental influences in the machine, such as cooling fluids. Power can be supplied using a cable (but this may not be possible in many machines), using batteries that have to be recharged at certain intervals or using other interfaces that the machine offers.

Since contamination in or on the optics of an optical 3D measurement instrument can significantly influence the measurement results, it has to be ensured that the device remains clean during use in the machine. To do so, different concepts can be used, such as a watertight housing for the sensor or specific regions in the machine where the sensor can be stored safely during the machining operation.

Table 3.2. Comparison of vibration sensitivity of different measurement technologies.

Technology	Vertical length of influence calculated by	Vertical length of influence/μm
Focus variation	FWHM	9.6
Confocal	FWHM	6.9
CSI	Coherence length	1.4

Overall, using an FV sensor for in-process measurement is possible but many additional factors need to be taken into account. The future demands of Industry 4.0 will drive the development in this direction (Gao *et al* 2019, Leach 2020).

3.3 Case studies

3.3.1 Case study for measurement of smooth surfaces

3.3.1.1 Measurement noise

Measurement noise M_N and instrument noise M_I are metrological characteristics defined in ISO 25178 part 600 (ISO 2019) and which will be described in detail in part 700 (ISO/CD 2019).

Traditionally, roughened material measures are used for the determination of noise with FV, since conventional FV requires a minimum amount of surface texture ($Ra \geqslant 9$ nm with $\lambda c = 2$ μm). For this case study, all measurements were obtained using a Bruker Alicona micro coordinate measurement machine (μCMM) equipped with several magnification (10×, 20× and 50×) objectives. The μCMM is shown in figure 3.14. Instead of a roughened surface, a flat mirror was used as the material measure and for all measurements, FV with a modulated source (see section 3.2.1, called 'SmartFlash 2.0' on this instrument) was used. The measurement noise and the instrument noise were determined using the subtraction method according to ISO/CD 25178 part 700, where the difference of two consecutive measurements was used to calculate the root mean square Sq and subsequently the measurement noise M_N. As stated in ISO/CD 25178 part 700, instrument noise was determined under the best environmental conditions, whereas the measurement noise has been determined under normal environmental conditions. The results of instrument and measurement noise evaluation are summarised in table 3.3.

There is a clear improvement of the measurement noise for each magnification compared to conventional FV (see Giusca *et al* 2014). The increase of the measurement noise with decreasing magnification is expected, but FV with a modulated source is approaching CSI-level measurement results, which are nearly constant for different magnification objectives. As expected, the instrument noise values are slightly better than the measurement noise results and demonstrate the improvement of noise behaviour by FV with the modulated source.

3.3.1.2 Measurement of optical components

The surface texture of high-quality optical components is in the range of a few nanometres and consequently, 3D measurement of such samples is not possible with conventional FV. Since FV instruments with modulated sources are able to measure optically smooth surfaces (see section 3.2.1), surface quality inspection of optical components becomes a field of application for these types of instruments. The range of applications for optical components, such as lenses, prisms, filters or mirrors is large, and some examples are medical devices, smartphone cameras and optical measurement instruments. In figure 3.15, various optical components used in smartphone cameras are shown.

Figure 3.14. Bruker Alicona μCMM equipped with a fully automatic tilt and rotation unit used for the measurements in the case studies for smooth surfaces and VFP. The μCMM is a fixed-bridge coordinate machine with a granite frame, where the x- and z-axes are constructed as a bridge over the moving y-axis.

Table 3.3. Overview of measurement and instrument noise results for FV with a modulated source.

Magnification/×	Numerical aperture	Measurement noise/nm	Instrument noise/nm
10	0.3	3.7	2.6
20	0.4	2.1	1.1
50	0.6	0.7	0.4

Surface imperfections on optical components, for example, scratches and digs, are a matter of concern for several reasons (Baker 2004, Gross *et al* 2012). They may degrade the functional performance. There is also a cosmetic interest: visible blemishes on high-quality optical parts will not be accepted by the customer, regardless of whether they impair performance or not. In high-power laser applications, surface imperfections can have severe impacts—strong light scattered

Figure 3.15. Optical components (typically lens diameters are around a few millimetres) in a smartphone camera. Image source: peshkov—stock.adobe.com.

from imperfections can cause a health risk and absorption of light by defects can provoke optical damage and eventually destroy the optics.

Quality control of optical surfaces is the scope of several specification standards, see ISO 10110 part 7 (ISO 2017a), ISO 14997 (ISO 2017b), ANSI/OEOSC OP1.002 (ANSI 2017) and US Standard MIL-PRF-13830B (US Department of Defense 1997). In these standards, surface imperfections are classified and test methods are described. Whereas in the MIL standard, surface imperfections are either scratches or digs, they are divided into more classes in the other standards. For example, in ISO 10110 part 7, surface imperfections are classified into long imperfections (scratches), localised surface imperfections (scuffs, pits, fixture marks, coating blemishes and adhered particles) and edge chips.

The test methods can be divided into two classes: dimensional methods and visibility methods. Dimensional methods determine the area affected by imperfections, for example, the diameter of a dig is compared to a calibrated reference. Visibility methods assess the visibility of imperfections, for example, the brightness of a scratch is compared to a calibrated reference. However, it is common to all methods that they are designed for manual inspection by human operators. Human evaluation of surface imperfections has numerous shortcomings, such as a lack of reproducibility, variation across operators, dependence on the training of the operator and the influence of human factors (for example, concentration, fatigue and mood). Moreover, human evaluation of optical surfaces is not well documented, for example, generally, there is no knowledge about the position of each defect. Finally, in addition to these drawbacks, there are two further issues that enhance the

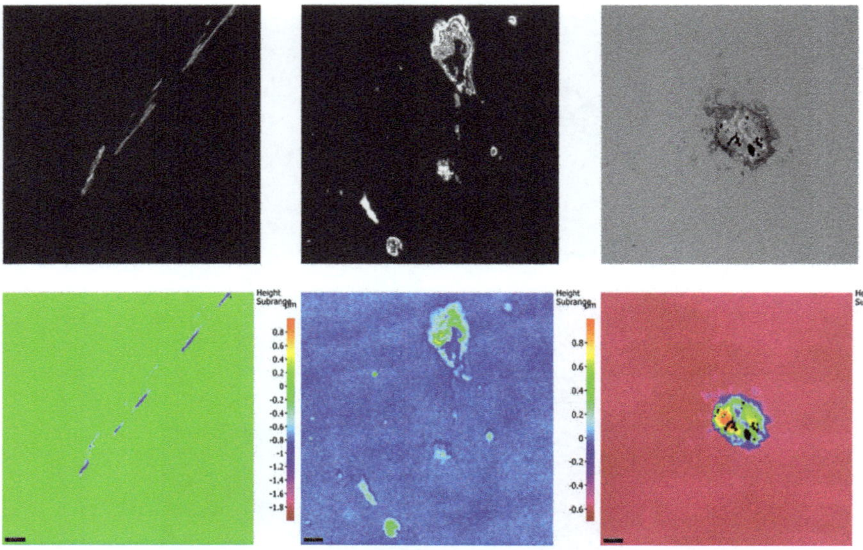

Figure 3.16. 3D measurements of a microscope cover glass (left column), a prism (centre column) and a plastic lens (right column). In the top line the true colour information and in the bottom line the height values are given.

need for an automated accurate and reproducible evaluation of optical surfaces. First, there is an ongoing miniaturisation of optical components, and this decrease of geometric dimensions is accompanied by much tighter tolerances on surface imperfections. Second, new production technologies for optical components have been developed. As an example, polymer lenses are produced by injection moulding (Lerma Valero 2020). Due to the small dimensions, the tight tolerances and the smooth surface conditions, quality control of the mould and the plastic optics is a difficult application.

There have already been some approaches for objective automated 2D inspection of optical surfaces (Etzold *et al* 2016, Neubecker and Hon 2016, Schöch *et al* 2018), and these approaches are mentioned in the version of ISO 14997 published in 2017 (ISO 2017b). Nevertheless, this document also describes dimensional and visibility methods for human evaluation of surface imperfections. The use of measurement instruments for automated quality control of optical surfaces is not within the scope of current standards.

Automated inspection with measurement instruments based on FV is a step towards an objective, accurate and reproducible quality control of optical components. Having an areal topography dataset of the optical component available can give valuable insights to the manufacturing process. In the case of plastic lenses produced by injection moulding, conclusions can be drawn about the condition of the mould.

The capability of FV instruments to measure optical components is shown in the following illustrations. A microscope cover glass, a prism and a plastic lens were measured on a Bruker Alicona μCMM using FV with modulated source. In figure 3.16, true colour images and 3D datasets in pseudo-colours of the three samples are shown. On the microscope cover glass, several thin surface imperfections (scuffs according to ISO 10110 part 7 (ISO 2017a)) are visible. Both of the other samples have small localised imperfections with height values above the optical surface.

A potential approach for the automated inspection of optical surfaces with FV instruments consists of the following steps:

- capture a course preview 2D dataset;
- localise surface imperfections in the 2D dataset;
- accurate 3D measurement of the relevant surface imperfections;
- evaluation of imperfections;
- digital documentation of imperfections, for example, single sample characteristics (number of scratches, size of each scratch, etc) as well as statistics over several samples; and
- evidence of consistency with relevant standard, for example, ISO 10110 part 7, and report results consistent with the notation used in the standard.

3.3.2 Case study for vertical focus probing

3.3.2.1 The use of vertical focus probing in high-precision optical dimensional metrology

VFP is mainly used in dimensional metrology where high-precision measurements of dimension, position and form are essential for quality assurance. The measurement of vertical walls with slope angles greater than 90° by lateral probing builds the basis for many high-precision coordinate measurement applications, for example, the measurement of lateral distances or the determination of the dimensional parameters of cylinders or spheres.

Optical measurement instruments use four- or five-axis configurations to measure the 3D form of a sample by measuring it from different orientations and then aligning the single datasets to a combined 3D dataset, for example, by using an additional rotation and tilt unit (Helmli 2011, Moroni *et al* 2014). Using VFP, manipulation of the sample can be avoided or considerably minimised by measuring all positions by lateral probing. Typically, such a procedure includes the following steps:

1. mount the sample;
2. measure the sample with VFP at several positions without manipulating the sample; and
3. merge the measured datasets to one 3D dataset.

In this case study, diameter measurements of common geometrical forms are presented—a cylindrical pin, a sphere and a hole—which occur frequently on various measurement artefacts. The calibrated samples (see table 3.4 for calibration details) were mounted so that the measured surfaces are parallel to the optical axis (see figure 3.17, top). The measured 3D datasets can be seen in figure 3.17 (bottom).

Table 3.4. Overview of the vertical focus probing measurement results on calibrated artefacts.

Artefact	Calibration institute	Unit	Calibrated value	Measured value	Deviation
Hole	METAS	mm	0.200 65 ± 0.000 13	0.200 568	0.000 082
Pin	DAkkS	mm	6.000 33 ± 0.000 50	6.000 92	0.000 59
Sphere	METAS	mm	0.999 72 ± 0.000 20	0.999 16	0.000 56

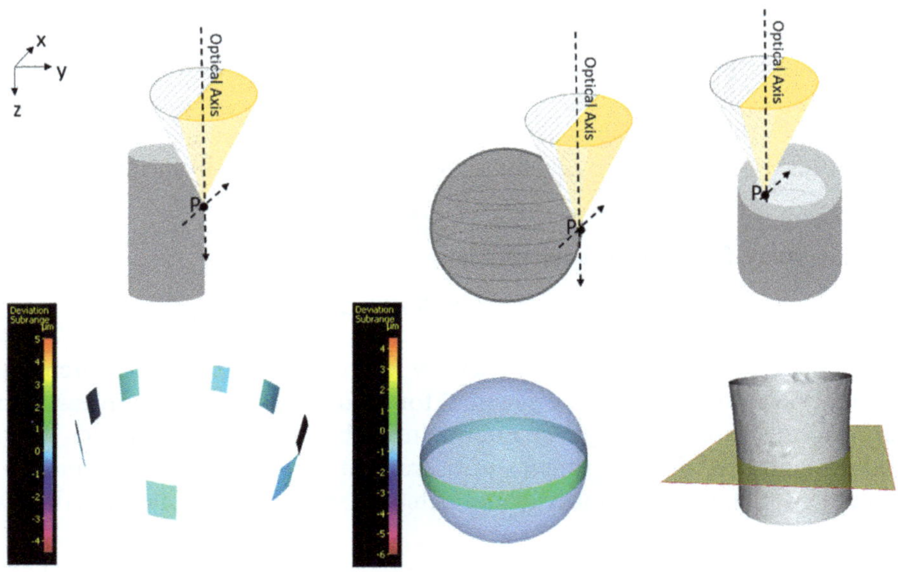

Figure 3.17. Top: Typical measurement set-up for VFP to measure the diameters of pins, spheres or holes. The optical axis is parallel to the measured surface. Bottom: Several VFP measurement results on calibrated artefacts are shown. The diameter of the artefacts was determined with least-squares fitting (see table 3.4).

The measured dataset of the pin consists of eight single measurements around its circumference. The measured 3D dataset of the equator ring of the sphere consists of sixteen single VFP measurements that are fused together into one 3D dataset. The measured diameter was calculated by fitting a least-squares cylinder or a least-squares sphere (as appropriate) for the 3D datasets. The results can be found in table 3.4. The deviation from the calibrated diameter is always less than 1 μm.

To measure lateral distances, a tactile measurement instrument would need to contact the sample laterally on both sides and measures the point-to-point distances. VFP allows the direct contactless measurement of distances between two sidewalls without the use of a rotation system.

In addition to the 3D form, the measurement of surface texture characteristics is essential for many applications in dimensional metrology. Figure 3.18 shows two measured surface datasets of the wall region of a cutting insert. The measurement on

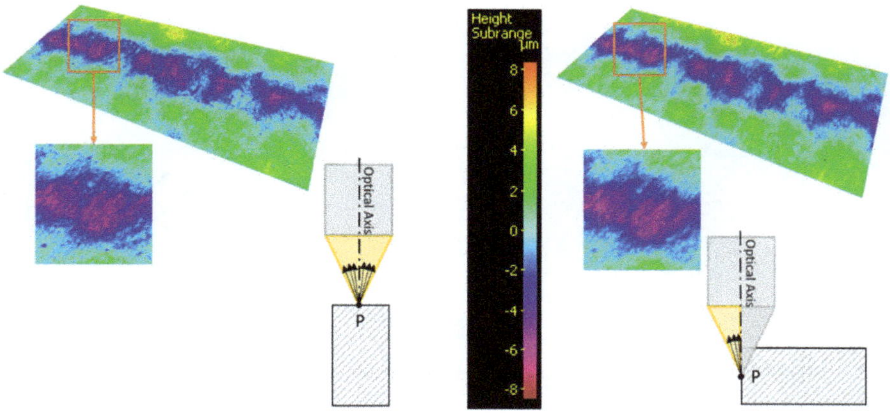

Figure 3.18. Left: FV measurement of a region. Right: Measurement of the same wall region of a cutting insert with VFP. The measurements show the same overall surface characteristics as well as very similar small surface structure features. The datasets are shown in pseudo-colour in a height range of -8 μm to 8 μm.

the right side was carried out using VFP and the wall region had a slope angle of 90°. For the measurement on the left side, the part was rotated by 90° and the same position was measured using FV. Both measurements were performed with a Bruker Alicona μCMM using a 20× objective. For improved comparability, the 3D datasets are shown in pseudo-colour. In both datasets, the same surface characteristics can be identified (see zoomed parts). A difference measurement shows that 96.9% of the surface points are within a 1 μm tolerance with a mean deviation of 0.06 μm.

3.3.2.2 Measurement of micro-holes in a fuel injector

The measurement of micro-holes is important in many industrial applications and challenging in dimensional metrology (Hansen *et al* 2006). There is only a limited number of measurement technologies that can measure such geometrical structures (for example, x-ray computed tomography (Carmignato *et al* 2017), fibre probes (Muralikrishnan *et al* 2006), FV with VFP). One important application in the automotive industry is the optimisation of the combustion process by measuring and analysing the geometry of holes in injector nozzles and determining its influence on the combustion process. In this case study, the measurement of the micro-holes in injection nozzles is discussed (see also Zangl *et al* 2018).

Figure 3.19 shows a schematic illustration of a fuel injection nozzle by cross-section, where the needle, the seat and two of several nozzle holes are presented. In general, the nozzle holes are cylindrical (hole inlet diameter = hole exit diameter) or cone-shaped (hole inlet diameter ≠ hole exit diameter). The relevant parameters are the form of the micro-hole as well as its orientation with respect to the part axis. In this application, the measured injection nozzle consists of eight injection holes with a diameter-to-depth ratio of 1:5. The diameter of the holes is typically between 80 μm and 600 μm. The orientation of a nozzle hole can be determined from the angle between the part axis and the hole axis.

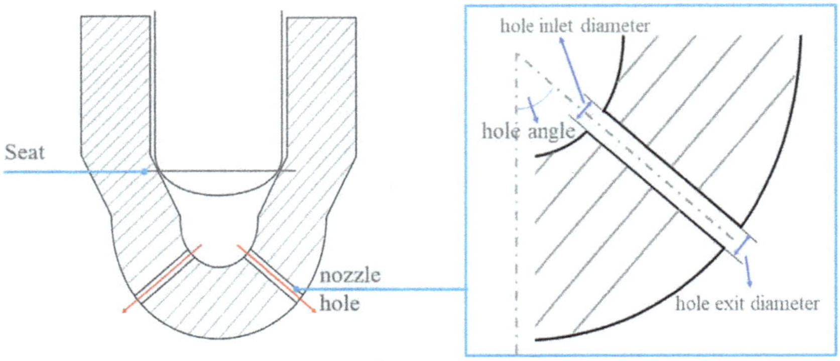

Figure 3.19. Cross-section of a fuel injector and measured injection hole geometry.

The measurements were carried out with a Bruker Alicona μCMM equipped with a high-precision rotation and tilt unit. The measurement of the injection nozzle and the micro-holes consists of the following steps (see figure 3.20):

1. mount the sample on the rotation unit;
2. measure the micro-holes using VFP;
3. measure the outer shell by a FV measurement; and
4. merge the measurements into one 3D dataset.

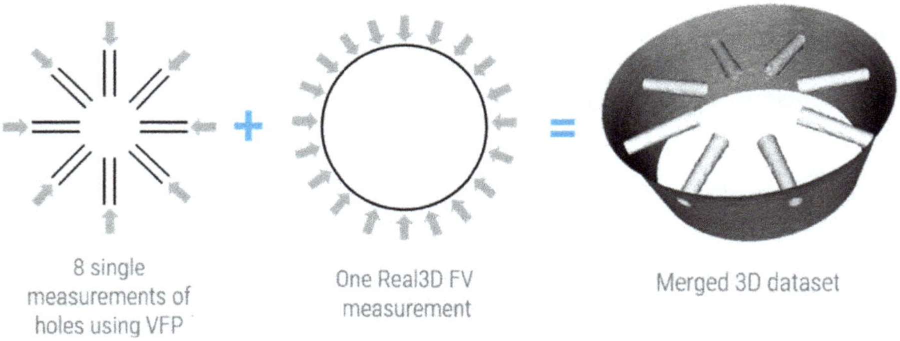

Figure 3.20. Procedure to measure an injection nozzle with several micro-holes.

The relevant geometry parameters were calculated according to the following procedure (see also figure 3.21):

1. determine the tool axis by a robust cone fit to the whole geometry;
2. determine the hole axis by a robust cone fit to each nozzle hole;
3. determine the position of the holes with respect to the tool axis by computing the angle between each hole axis and tool axis; and
4. determine the inner and outlet hole diameters with robust circle fits into the cross-section of the inlet and outlet of the bore, using standard Gaussian fits with 3σ and equally weighted measured points.

Figure 3.21 shows the measured 3D dataset as well as the single steps in the geometry evaluation. An overview of the evaluation of the nozzle hole orientations

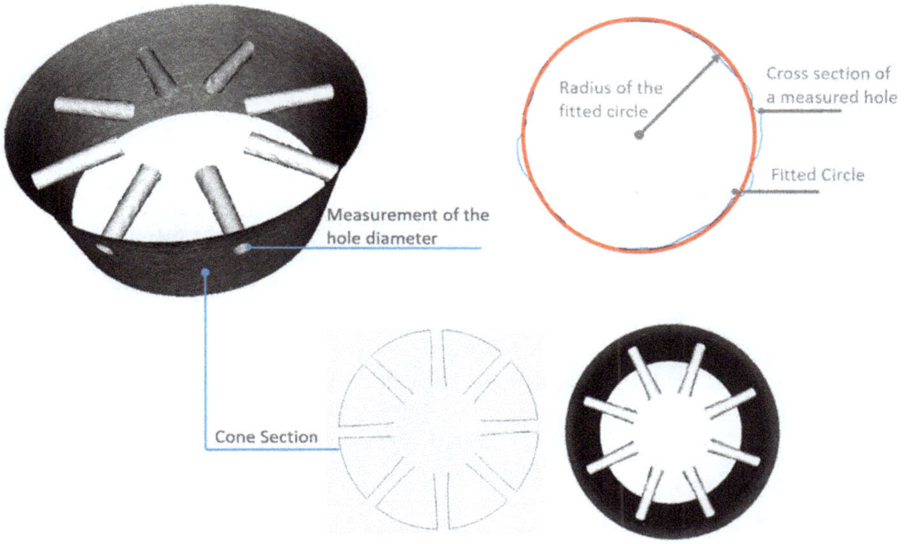

Figure 3.21. Evaluation of the nozzle hole geometry and the geometric correlations of multiple holes.

with respect to the tool axis is given. Additionally, the cross-section through the injection holes shows their conical shape which can also be verified by evaluation of their geometry. The evaluation of the first hole results in a hole inlet diameter of 0.145 mm, a hole outlet diameter of 0.126 mm and a cone angle of 3.038°, which demonstrate that the hole can be measured by VFP even though it has conical form and increases its diameter towards the inlet.

3.3.3 Case study for in-process surface measurement

As described in section 3.2.3, advanced manufacturing processes need to be highly efficient and accurate, as well as containing low error rates. In-process metrology is one approach to fulfil these requirements. In this section, the use of FV for on-machine surface measurements directly in a turning machine (lathe) is presented. The 'collaborative' FV measurement systems (Danzl *et al* 2017, Riedl *et al* 2019) are based on the combination of a collaborative six-axis robot (cobots) and a robust optical 3D measurement sensor. The cobots have interfaces (for example, TCP/IP, Modbus/TCP or Anybus) for the connection to existing production systems. This facilitates the communication between the individual machines and allows the implementation of a self-controlling and a self-correcting production. The system shown in figure 3.22 allows for easy course positioning with two-handle bars, fine positioning with integrated joysticks, fully automated measurement and subsequent measurement reporting including go/no-go statuses.

One field of application for the cobot is in the development of new cutting materials. Typically, new cutting tools are developed by monitoring the cutting tool during its lifetime and trying to understand the dominant wear modes and underlying wear mechanisms. The knowledge gained is used for the creation of new cutting tools. Therefore, it is important to measure the wear at intervals along the

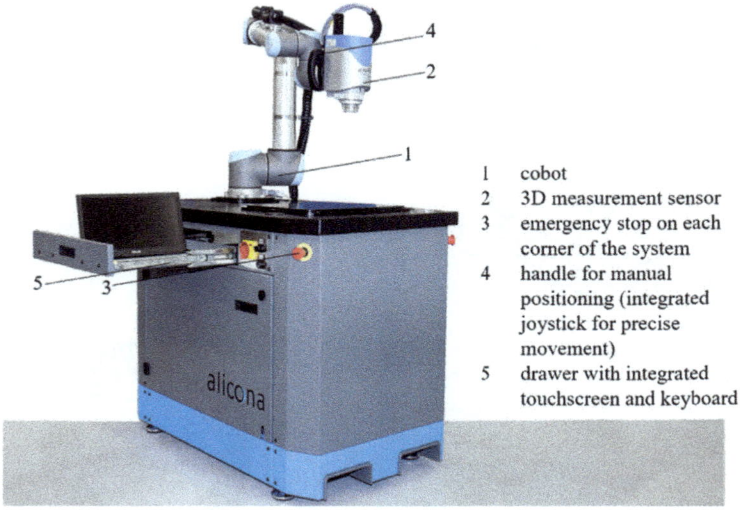

Figure 3.22. Collaborative measurement system for quality control in the production line.

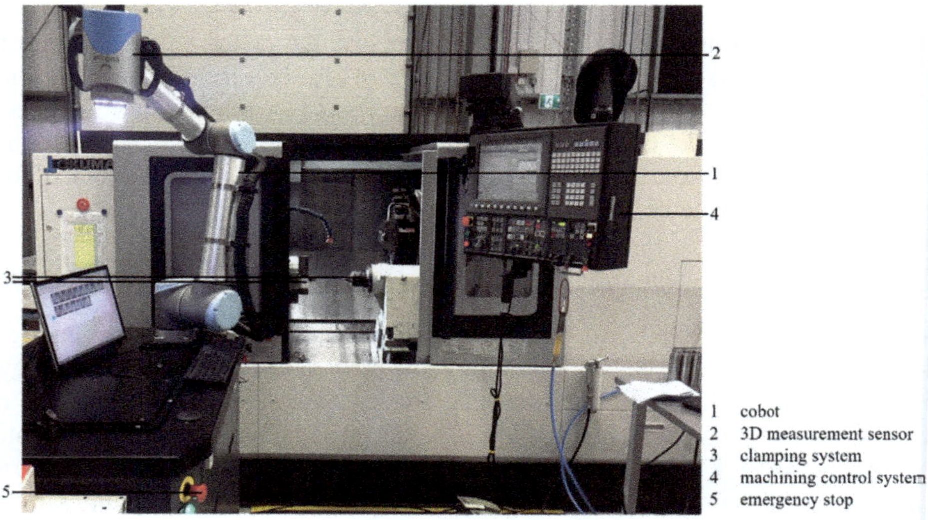

Figure 3.23. On-machine measurement directly in the lathe. In the front you see the cobot and in the back the lathe is visible. Image source: Element Six.

cutting tool's life. This periodic inspection is a limiting factor for the development of new cutting tool materials. Traditionally, an operator takes the tool out of the machine, brings it to a laboratory to quantify the wear, and then returns it to the machine. This routine is repeated several times a day for each machine and clearly impacts productivity.

To improve productivity and shorten the development time of new high-performance cutting tool materials, companies can use cobot measurement systems directly in the lathe and automate the periodic inspection process. In figure 3.23, this integrated production strategy is illustrated. The implementation of an automated test series enables the automatic measurement of the wear status of the cutting tool as well as of the workpiece surface condition. The following process is repeated periodically and replaces the above-described traditional steps performed by an operator. The on-machine measurements are started by a central control system. At a defined point, the lathe stops and a signal sets the cobot in motion. The cobot arm equipped with a 3D measuring sensor is automatically manipulated into the lathe and first measures the wear parameters of the cutting tool (see figure 3.24). Then, the sensor moves on to the workpiece and measures surface texture parameters to verify the surface quality. When all measurements are finished, the cobot arm returns into its original position. In figure 3.25, the measurement result of a used cutting tool is shown. The left image depicts the 3D dataset, and the differences to an unused reference insert are shown in the right image. In this way, the task of generating the data has been removed from the operator. Now, the operator can evaluate the measurement, for example, by analysing wear parameters, such as flank, crater and notch wear that are defined and described in detail in ISO 3685 (ISO 1993) and ISO 8688 part 1 (ISO 1989). Based on the measurement results, it may be necessary to

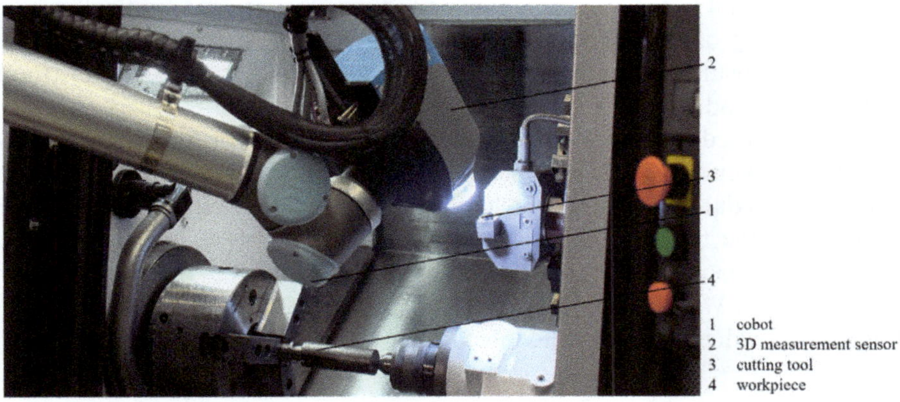

Figure 3.24. Measurement of the cutting tool directly in the lathe (on-machine measurement). Image source: Element Six.

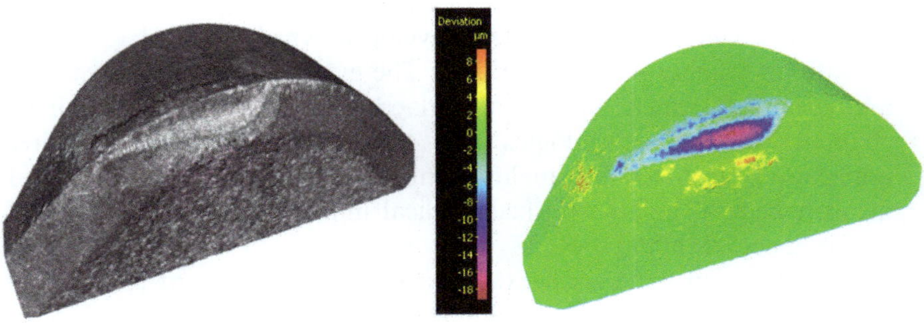

Figure 3.25. Left image: 3D dataset of a used cutting insert. Right image: Differences to an unused reference cutting insert. Image source: Element Six.

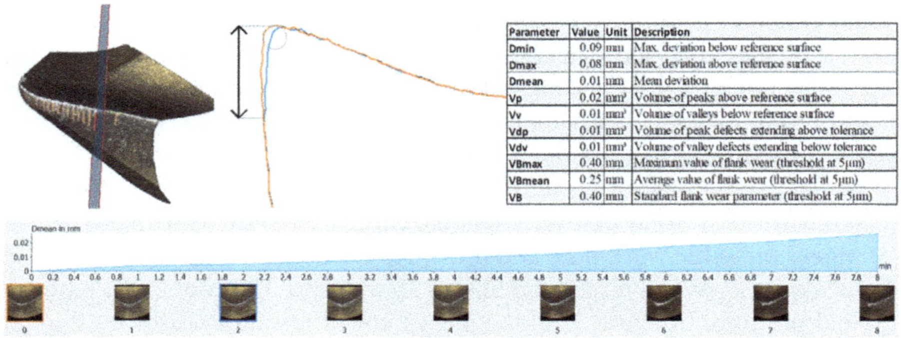

Parameter	Value	Unit	Description
Dmin	0.09	mm	Max. deviation below reference surface
Dmax	0.08	mm	Max. deviation above reference surface
Dmean	0.01	mm	Mean deviation
Vp	0.02	mm³	Volume of peaks above reference surface
Vv	0.01	mm³	Volume of valleys below reference surface
Vdp	0.01	mm³	Volume of peak defects extending above tolerance
Vdv	0.01	mm³	Volume of valley defects extending below tolerance
VBmax	0.40	mm	Maximum value of flank wear (threshold at 5μm)
VBmean	0.25	mm	Average value of flank wear (threshold at 5μm)
VB	0.40	mm	Standard flank wear parameter (threshold at 5μm)

Figure 3.26. Wear measurement of a time series of 3D datasets. The wear parameters are calculated for each cutting plane position.

change the machine parameters before starting the lathe again and continuing with the machining of the workpiece.

The next step in the automation process is the automatic correction of machine parameters based on the measurement results. The cobot transmits measured values or a go/no-go signal to the lathe and machine parameters are automatically changed before the testing of new cutting tool materials continues. It is expected that the production of unsuitable parts is considerably reduced this way—ideally, the first part is already produced as a good part.

Using specialised wear analysis software, an example of an automated wear measurement on cutting tools is presented in figure 3.26. The tool wear is evaluated on a series of 3D datasets measured over a defined period of time. The first dataset in the timeline is the reference dataset and wear parameters, such as flank wear or notch wear for each dataset in the timeline, are measured with respect to the reference dataset. Wear parameters for each cutting plane position as well as mean and maximum values are calculated according to ISO 8688 part 1.

3.4 Conclusions and outlook

In this chapter, three new measurement approaches based on FV have been presented.

- Conventional FV instruments require a minimum degree of surface texture but new technologies have been developed which allow FV to measure smooth and highly polished surfaces. The evaluation of measurement noise and the measurement of optical surfaces were discussed in detail in this chapter. The new technologies have opened up challenging new areas of application, such as the quality control of optical components and the measurement of highly polished medical implants, cutting tools and high-performance stamping tools.
- Vertical focus probing is an extension to the measurement principle of FV and enables the measurement of slope angles greater than 87° directly and without the use of any articulating system. It is an optical method and allows for a fast, contactless and area-based measurement. The presented examples

show that vertical focus probing opens up new possibilities regarding coordinate measurement applications (lateral probing, measurement of vertical walls and holes, measurement of complex geometries) and provides the opportunity to close the gap between tactile and optical coordinate measurement instruments. Bidirectional measurements according to ISO 10360 part 8 (ISO 2013) with optical measurement instruments based on vertical focus probing are now possible.

- Modern manufacturing processes have demanding requirements regarding accuracy, efficiency, tolerances and low error rates. A promising approach to address these requirements is in-process quality control in the production line. A measurement instrument, which is suitable for this task, must fulfil numerous additional requirements compared to its use in a laboratory environment, such as high-speed measurement, robustness to vibrations and robustness to temperature variations. When it comes to vibration, research has shown that FV is a robust technology. The use of a collaborative measurement system, which uses FV as the measurement principle, for on-machine measurements directly in a lathe or milling machine clearly demonstrates the benefits of in-process surface metrology.

References

Akcay C, Parrein P and Rolland J P 2002 Estimation of longitudinal resolution in optical coherence imaging *Appl. Opt.* **41** 5256–62

ANSI 2017 *Optics and Electro-optical Instruments—Optical Elements and Assemblies—Surface Imperfections ANSI/OEOSC OP1.002* (Rochester, NY: American National Standards Institute)

Artigas R 2011 Imaging confocal microscopy *Optical Measurement of Surface Topography* ed R K Leach (Berlin: Springer)

Baker L R 2004 *Metrics for High-quality Specular Surfaces* (Bellingham, WA: SPIE Press)

Bermudez C, Martinize P, Cadevall C and Artigas R 2019 Active illumination focus variation *Proc. SPIE* **11056** 110560W

Carmignato S, Dewulf W and Leach R K 2017 *Industrial X-ray Computed Tomography* (Berlin: Springer)

Coveney T 2014 Dimensional measurement using vision systems *NPL Measurement Good Practice Guide* GPG39

Danzl R, Helmli F and Scherer S 2011 Focus variation—a robust technology for high resolution optical 3D surface metrology *J. Mech. Eng.* **57** 245–56

Danzl R, Lankmair T, Calvez A and Helmli F 2017 Robot solutions for automated 3D surface measurement in production *Proc. 18th Int. Congr. Metrol.* 15002

Etzold F, Kiefhaber D, Warken A F, Würtz P, Jon J and Asfour J-M 2016 A novel approach towards standardizing surface quality inspection *Proc. SPIE* **10009** 1000908

Gao W, Haitjema H, Fang F Z, Leach R K, Cheung C F, Savio E and Linares J M 2019 On-machine and in-process surface metrology for precision manufacturing *Ann. CIRP* **68** 843–66

Giusca C L, Claverley J D, Sun W, Leach R K, Helmli F and Chavigner M P J 2014 Practical estimation of measurement noise and flatness deviation on focus variation microscopes *Ann. CIRP* **63** 545–8

Gross H, Dörband B and Müller H 2012 Testing texture and imperfections of optical surfaces *Metrology of Optical Components and Systems Handbook of Optical Systems* vol 5 (Weinheim: Wiley) pp 785–838

Hansen H N, Carneiro K, Haitjema H and De Chiffre L 2006 Dimensional micro and nano metrology *Ann. CIRP* **55** 721–43

Hecht E 2012 *Optics* (Harlow: Pearson)

Helmli F 2011 Focus variation instruments *Optical Measurement of Surface Topography* ed R K Leach (Berlin: Springer)

ISO 8688 part 1 1989 *Tool Life Testing in Milling—Part 1: Face milling* (Geneva: International Organization for Standardization)

ISO 3685 1993 *Tool-life Testing with Single-point Turning Tools—ISO 3685* (Geneva International Organization for Standardization)

ISO 10360 part 8 2013 *Geometrical Product Specifications (GPS)—Acceptance and Reverification Tests for Coordinate Measuring Systems (CMS)—Part 8: CMMs with optical distance sensors* (Geneva: International Organization for Standardization)

ISO 25178 part 606 2015 *Geometrical Product Specification (GPS)—Surface Texture: Areal—Part 606: Nominal characteristics of non-contact (focus variation) instruments* (Geneva: International Organization for Standardization)

ISO 10110 part 7 2017a *Optics and Photonics—Preparation of Drawings for Optical Elements and Systems—Part 7: Surface imperfections* (Geneva: International Organization for Standardization)

ISO 144997 2017b *Optics and Photonics—Test Methods for Surface Imperfections of Optical Elements* (Geneva: International Organization for Standardization)

ISO 25178 part 600 2019 *Geometrical Product Specification (GPS)—Surface Texture: Areal—Part 600: Metrological characteristics for areal topography measuring methods* (Geneva: International Organization for Standardization)

ISO/DIS 25178 part 700 2020 *Geometrical Product Specification (GPS)—Surface Texture: Areal—Part 700: Calibration, adjustment and verification of areal topography measuring instruments* (Geneva: International Organization for Standardization)

Leach R K 2011 *Optical Measurement of Surface Topography* (Berlin: Springer)

Leach R K 2014 *Fundamental Principles of Engineering Nanometrology* 2nd edn (Amsterdam: Elsevier)

Leach R K 2020 *Integrated Metrology 10-year Roadmap for Advanced Manufacturing* (HVM Catapult) https://metrology.news/integrated-metrology-a-10-year-roadmap-for-advanced-manufacturing/

Lerma Valero J R 2020 *Plastics Injection Molding—Scientific Molding, Recommendations, and Best Practices* (München: Carl Hanser)

Matilla A, Mariné J, Pérez J, Cadevall C and Artigas R 2016 Three-dimensional measurement of technical surfaces with simultaneous scan of confocal and focus variation *Proc. SPIE* **9890** 98900B

Moroni G, Petrò S and Syam W P 2014 Four-axis micro measuring systems performance verification *Ann. CIRP* **63** 485–8

Muralikrishnan B, Stone J A and Stoup J R 2006 Fiber deflection probe for small hole metrology *Precis. Eng.* **30** 154–64

Neubecker R and Hon J E 2016 Automatic inspection for surface imperfections: requirements, potentials and limits *Proc. SPIE* **10009** 1000907

Nikolaev N, Petzing J and Coupland J M 2016 Focus variation microscope: linear theory and surface tilt sensitivity *Appl. Opt.* **55** 3555–65

Noguchi M and Nayar S K 1994 Microscope shape from focus using active illumination *Proc. 12th ICPR* vol 1 pp 147–52

Pertuz S, Puig D and Garcia M A 2013 Analysis of focus measure operators for shape-from-focus *Pattern Recognit.* **46** 1415–32

Piska M and Metelkova J 2014 On the comparison of contact and non-contact evaluations of a machined surface *MM Sci. J.* 476–80

Riedl M, Danzl R, Felberbauer M and Lankmair T 2019 High-resolution geometry measurement with a collaborative robot *MM Sci. J.* Special Issue HSM 2019–57

Schöch A, Perez P, Linz-Dittrich S, Bach C and Ziolek C 2018 Automating the surface inspection on small customer-specific optical elements *Proc. SPIE* **10679** 1067915

Schuth M and Buerakov W 2017 *Handbuch Optische Messtechnik: Praktische Anwendungen für Entwicklung, Versuch, Fertigung und Qualitätssicherung* (Munich: Carl Hanser)

Tian Y, Weckenmann A, Hausotte T, Schuler A and He B 2013 Measurement strategies in optical 3-D surface measurement with focus variation *Proc. ISMQC* **40**

US Department of Defense 1997 Optical components for fire control instruments: general specification governing the manufacture, assemble, and inspection *Performance Specification* MIL-PRF-13830B

VDI 2015 *Measuring Rooms—Classification and Characteristics—Planning and Execution—VDI 2627-1* (Düsseldorf: Verlag des Vereins Deutscher Ingenieure)

von Helmholtz H L F 1924 *Helmholtz's Treatise on Physiological Optics* (New York: Optical Society of America)

Wang J, Leach R K and Jiang X 2015 Review of the mathematical foundations of data fusion techniques in surface metrology *Surf. Topog. Metrol. Proper.* **3** 123001

Zangl K, Danzl R, Helmli F and Prantl M 2018 Highly accurate optical μCMM for measurement of micro holes *Proc. CIRP* **75** 397–402

Zangl K, Danzl R, Muraus U, Helmli F and Prantl M 2019 Vertical focus probing for high-precision optical dimensional metrology *Proc. ISMTII*

Chapter 4

Imaging confocal microscopy

Roger Artigas

The three most common optical surface topography measuring instruments are coherence scanning interferometry (see chapter 2), focus variation microscopy (see chapter 3) and imaging confocal microscopy. Any surface can be measured with one of the three techniques with different results depending on the surface interaction with light, the optical characteristics, and the metrological performance of each technology. Confocal microscopy benefits from a high numerical aperture and high magnification objectives. Three-dimensional surface texture can be reconstructed on a pixel-by-pixel basis with nanometre resolution. This chapter will review imaging confocal microscope technologies for the areal measurement of surfaces, and their calibration and performance characteristics. Advanced methods for topography information recovery will also be reviewed: structured illumination microscopy, active illumination focus variation and the combination of both techniques to provide improved performance and results.

4.1 Introduction

According to the ISO 25178 part 607 (ISO 2019), a confocal microscope for three-dimensional (3D) measurement is a system in which the measurement method comprises the localisation of optically sectioned images during an axial scan through the focus of a microscope's objective, providing a means to determine an areal surface topography image. The confocal microscope produces optically sectioned images by restricting the illumination onto the sample and through the detection system by means of a pattern, scanning this pattern in-plane to fill the image. The illumination and detection patterns could be one or several points, slits or any type of pattern that effectively reduces the illuminated area of the surface. The geometry of these patterns influences the properties of the sectioned images and has a direct influence on the metrological characteristics of the instrument. The difference between a confocal point sensor and a confocal microscope is defined by the in-plane scanning scheme. In the confocal microscope, one or multiple parallel working

doi:10.1088/978-0-7503-2528-8ch4

light paths scan the surface by means of several optical and mechanical elements. Section 4.2 discusses different opto-mechanical arrangements to acquire confocal images. Independent of their opto-mechanical arrangement, all confocal microscopes need to perform an axial scan through the focus of the objective, recovering for each pixel a signal known as the axial response. A confocal peak location algorithm is used to estimate the maximum signal position pixel-by-pixel thus recovering the areal image. Ideally, the areal topography of an ideal flat mirror placed perpendicular to the optical axis of a confocal microscope should give a perfectly flat response, but optical aberrations of the microscope, such as Petzval and field curvature, mean that the response is not flat. This flatness error can be adjusted using a flat single surface mirror with a flatness better than $\lambda/10$ and sub-nanometre surface texture. However, a residual flatness error may appear after the adjustment process due to non-linear effects—this issue will be analysed in section 4.2.2.3, but there is still not a clear understanding of its cause and there is no universal proposal for its adjustment.

This chapter will cover two particular cases of areal surface measurement technologies using the optical sectioning property: structured illumination microscopy (SIM) and active illumination focus variation, in sections 4.3 and 4.5, respectively. The SIM microscope is used mainly in life science applications to minimise the photobleaching toxicity while providing optically sectioned images, similar to those from a laser scanning microscope. On highly scattering surfaces or deeply focused structures, laser scanning provides higher contrast due to less crosstalk between the illumination and detection patterns. In contrast, for single reflective layers, such as those found in many engineering surfaces, there is no difference between both types of optically sectioned images, and an SIM microscope can be used for 3D measurements with much higher efficiency, speed and simpler hardware. Although focus variation is not a confocal arrangement, active illumination focus variation shares many properties with confocal microscopy: there is an active pattern that restricts the amount of illumination incident on the surface but the detection is performed by a focus operator. The result is an optically sectioned image similar to that from a confocal microscope, but with an inherent loss of lateral resolution due to the nature of the focus operator. Even with the loss of lateral resolution, there is still a large number of applications for this hybrid approach.

Note that much of the basics of imaging confocal microscopy can be found in Artigas (2011) and Liu and Tan (2016) and will not be repeated in detail here.

4.2 Confocal microscopy in the ISO 25178 framework

Depending on the in-plane scanning scheme, a confocal microscope can fall into one of the three following categories: laser scanning, disc scanning and programmable array microscopes (ISO 25178 part 607 (ISO 2019, Artigas 2011)). With a laser scanning confocal microscope, the illumination and observation patterns are composed of single pinholes and a laser is used to illuminate them. The pinhole image is scanned over the surface by means of scanning mirrors. A laser scanning system has the advantage of having the highest contrast optical section at the cost of using a coherent light source that can lead to issues due to speckle on rough surfaces. In disc scanning, the pinholes are arranged on a disc that fills the surface under inspection during its rotation. A disc scanning confocal microscope can provide hundreds of images every second, but typically has low light efficiency. To minimise light loss, larger pinholes and polarisation components are used. Programmable array microscopes are also known as microdisplay scanning systems. They use a reflective type microdisplay to dynamically change the illumination and detection pattern to give the highest optically sectioned capability or to maximise light efficiency, depending on the numerical aperture and the needs of the surface under inspection.

4.2.1 Laser scan, disc scan and microdisplay scan confocal microscopes

4.2.1.1 Laser scanning confocal microscope

A Laser scanning confocal microscope (LSCM) is based on the original concept of optically sectioning microscopy invented by Marvin Minsky in 1957. The optical setup of an LSCM in reflection mode is shown in figure 4.1: a laser light source is directed to a pinhole which is located at the field diaphragm position of a conventional microscope. The size of the pinhole is chosen to form an image at the focal plane of a high numerical aperture objective as small as possible, typically close to the diffraction limit. The light reflected or scattered by the surface under inspection passes back through the microscope objective and its image is formed on a conjugate plane where a second pinhole and a photodetector are placed. This second pinhole is acting as a light sectioning filter, since the light reflected on the surface from planes above and below the focal plane of the objective are forming an image in front or at the back of this pinhole, having a much larger size at the confocal aperture plane. The photodetector is registering high signal values for those regions of the surface lying in the focus of the objective and low signal for the rest.

To measure the height value of a surface an LSCM performs a vertical scan by shifting the relative position of the focus of the objective to the surface under inspection. The most used technique is to displace the microscope objective by means of a piezo-actuated linear stage. Because of the single point nature, an additional point by point XY scanning is necessary for the measurement of the three-dimensional components of a surface.

The single point setup can be extended to a line confocal scanning (Sheppard *et al* 1988) by projecting a slit onto the surface and detecting the signal along several points on a pixelated detector. The benefit is a much higher measurement speed at the cost of crosstalk between pixels.

A typical approach to measure 3D surfaces with an LSCM is by acquiring 2D confocal images at several height positions. Each confocal image is acquired by scanning the image of the pinhole (James and Pawley 2010) on the objective's focusing plane. Several configurations have been proposed for the last 40 years, the use of a couple of scanning mirrors being the most typical. A laser light source is needed on an LSCM due to the fact that every point on the confocal image has been illuminated by a tiny amount of time: a 1 Mpixel image at 1 frame/second means that every pixel has

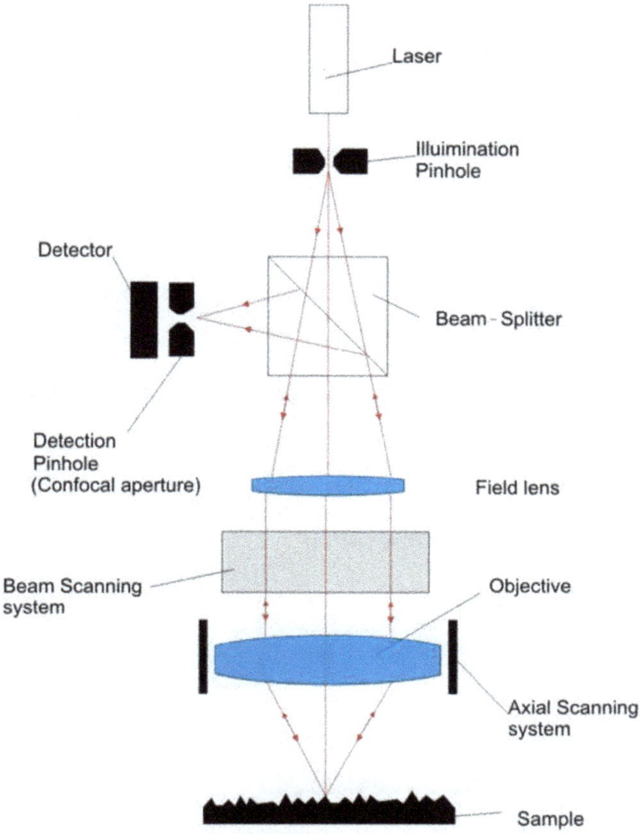

Figure 4.1. Typical arrangement of an LSCM.

been exposed for only 1 micro-second. Higher frame rate or even higher pixel count decreases the exposure time, and thus a high-density light source is needed.

4.2.1.2 Disc scanning confocal microscope configuration

In comparison to LSCM, a Disc scanning confocal microscope (DSCM) uses a set of several pinholes to illuminate and filter the optical section at many points simultaneously. Figure 4.2 shows the typical optical layout of this kind of confocal arrangement: a light source is expanded, and its light directed onto a disc. The disc has a set multiple pinholes on a spiral pattern. Each pinhole is imaged onto the surface and the light reflected or scattered goes back to its original pinhole that is acting as a focus filter itself. The disc plane is imaged onto a camera that is simultaneously imaging the surface. By rotating the disc at high speed, an optical section of the surface is imaged on the camera.

The main advantage of a DSCM is it high speed imaging capabilities which can achieve up to 1000 fps on a very special condition. Most of the light reflected onto the disc surface is reflected and directed to the imaging camera creating a background signal. Even with the use of an antireflection coating, this background could

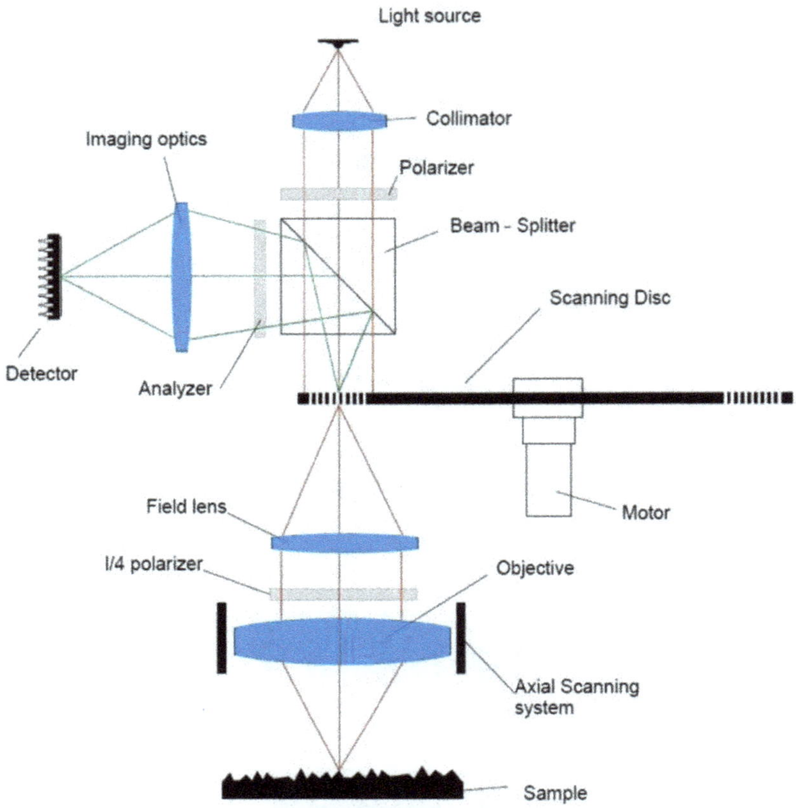

Figure 4.2. Typical arrangement of a DSCM.

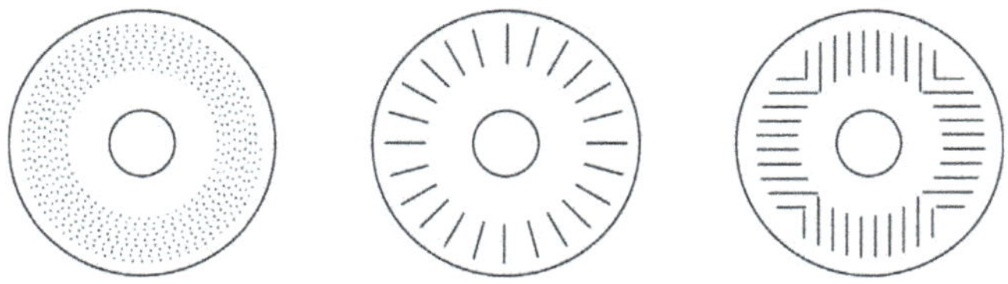

Figure 4.3. Three different disc patterns of a DSCM. Left to right: Nipkow disc, parallel slits and point rotating slits.

be larger than the signal reflected by the surface. A set of several polarizers and retarders are used to suppress the disc reflection without affecting the light reflected on the surface under inspection. The use of polarizing components lowers the light efficiency that combined with the low percentage of pinhole area on this disc makes a DSCM very poor in efficiency.

To increase the light transmission of the disc several patterns has been used. Figure 4.3 shows some of the typical arrangements: Nipkow, parallel slits and point rotating slits.

4.2.1.3 Programmable array scanning confocal microscope configuration

Programmable array (scanning confocal) microscopes (PAMs) are a digitally addressable optical arrangement that combines the benefits of an LSCM and DSCM on an electronically configurable scheme (Artigas *et al* 2004). Figure 4.4 shows the optical layout: a light source is expanded and directed to a microdisplay. The microdisplay is made of several pixels that can be turned on or off and thus directing the light to the optical axis or not. The microdisplay plane is imaged on the surface under inspection that is simultaneously imaged onto a camera. A very simple example to better understand its working principle consists of turning ON only one pixel. In this situation, only one point of the surface will be illuminated. If the surface on that point is in focus, most of the light will reach one pixel of the camera, while if that point is not in focus, the light will be spread onto many pixels, and the previous pixels will receive much lower signal. This example can be extended to a set of several pixels, slits, or any other pattern that restricts the amount of light onto the surface. A confocal image is reconstructed by shifting the pattern and recording images of the camera (Bitte *et al* 2000).

There are several microdisplay technologies on the market, but the most used on a PAM microscope are the ones having binary nature (only ON/OFF states). These

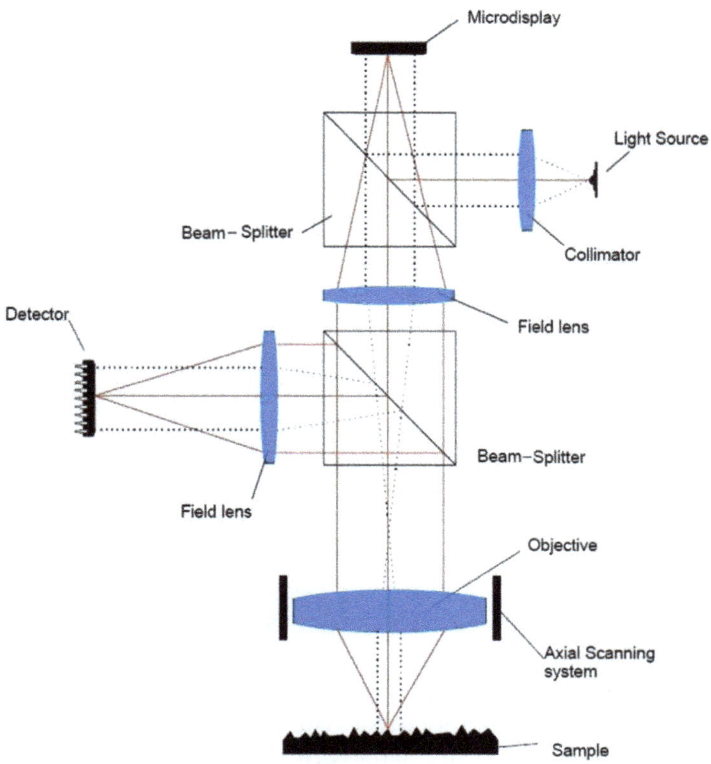

Figure 4.4. Typical arrangement of a PAM confocal microscope.

are LCoS and DLP, both being very similar in terms of pixel size, and achievable framerate.

4.2.2 Calibration, adjustment, performance specifications and influence factors for imaging confocal microscopes

According to ISO/CD 25178 part 700 (ISO 2020) the calibration of an areal measurement instrument refers to a series of operations required to establish the contribution of the metrological characteristics to the measurement uncertainty associated with the instrument's measurements. An instrument is calibrated and adjusted using traceable material measures, such as step heights and flatness specimens (see Leach *et al* 2015). The system can be adjusted, and the measurement quantity verified to match the calibrated value. A set of further tests are executed to check the performance specification and determine the instrument characteristics, such as accuracy, repeatability and reproducibility.

Despite the fact that the in-house set-up procedure of an areal instrument depends on the specific implementation of every manufacturer, there are a set of steps in the ISO 25178 standards for any microscope-based system that will not be discussed in detail here (but see Leach *et al* 2015). Performance specification will be discussed in the following sections. An important influence affecting confocal microscopes is the *residual* flatness error that comes from a non-perfect calibration and adjustment process. This error is still not well understood and to date there are no commercial instruments applying a correction. The *residual* flatness error will be discussed in detail in section 4.2.2.3.

4.2.2.1 Calibration and adjustment

There are two primary systematic errors in imaging confocal microscopes (ICMs) that need to be calibrated and adjusted. These are the flatness error and the amplification coefficient.

The flatness error is a consequence of aberrations of the optical components comprising the microscope. These components include the microscope objective itself, but also the illumination distribution, the field lenses and beam-splitters. The optical aberrations could have influences, such as distortions perpendicular and parallel to the measuring axis. In particular, Petzval, field curvature distortion and spherical aberration can cause the topography of a flat sample not to be measured flat, but rather to have a shape dictated by the combined effect of all the sources of aberration. The flatness error is calibrated by measuring a flat sample, such as a mirror with better than $\lambda/10$ flatness and sub-nanometre surface texture. Both characteristics of the calibration mirror are much less than the sensitivity of a high-level ICM. The adjustment process for the flatness error consists in the subtraction of the calibrated areal topography to ensure a white noise result. Figure 4.5 shows the calibrated and adjusted (left and right) flatness error. From figure 4.5, it can be seen that the flatness error does not have a symmetrical shape. There could be many reasons for this shape, such as the flatness of the beam-splitter components, lens stress during the manufacturing process, dominant directions due to scanning

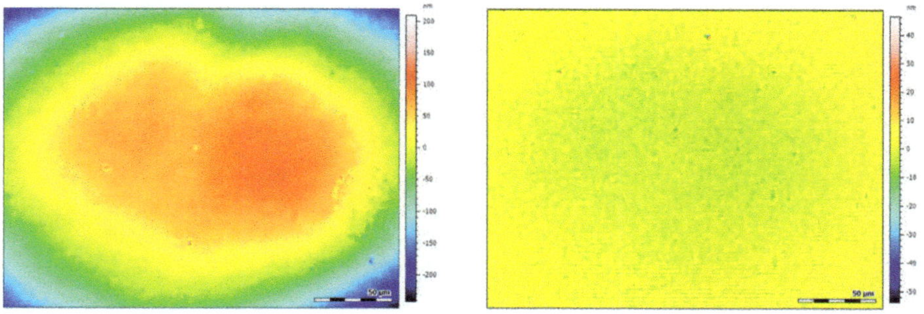

Figure 4.5. Calibrated (left) and adjusted (right) flatness error for a 50×, 0.8 NA objective.

mechanisms, and more. The adjustment process should in principle leave a flat noisy topography, but in many situations this is not the case, and low spatial frequency components appear, causing what is known as residual flatness error.

Most commercial ICMs use a mechanical vertical scanner, such as a linear stage, in both open-loop and closed-loop configurations with linear encoder scales. Higher performance closed-loop scanners, such as piezoelectric actuated stages with capacitive or piezoresistive sensors, are also used although they have a limited travel range. Independent of the type of scanner used, there will be two characteristics affecting the ICM: the parallelism between the measuring optical axis and the scanner, known as Abbe error (Leach 2015a), and the pitch error of the scanner. Both errors are systematic and influence the measured quantities by a multiplicative number. The calibration process consists of the measurement of a traceable step height artefact. The adjustment process is performed by multiplying the measured value by the amplification coefficient. A verification step is further performed to ensure that the system is performing to specification.

4.2.2.2 Performance specifications

Most of the metrological characteristics of an ICM are affected by the numerical aperture (NA) of the objective and the scanning mechanism to perform the vertical scan. Note that the lateral resolution of the 2D optical section is diffraction limited and shorter wavelengths produce sharper images, but the ability to recover the 3D information relies on the quality of the axial response, which is directly related to the appearance of speckle in the out-of-focus regions (Artigas *et al* 1999). Confocal systems using lasers as the light source, are prone to create out-of-focus speckle that on certain surfaces can result in higher intensities of the axial response out of the focal region and spikes in the topography map. An incoherent or partially coherence light source such as an LED provides cleaner images.

The NA of the objective influences the *measurement and instrument noise* (N_m) (see Leach *et al* 2015b for definitions), as well as the measurement repeatability. The *flatness deviation* (Z_{FLT}) (see Leach *et al* 2015b for definition) is also related to the NA (see section 4.2.2.3).

Figure 4.6 shows two of the most relevant influence factors of an areal measurement system: the instrument noise and the maximum measurable slope on optically

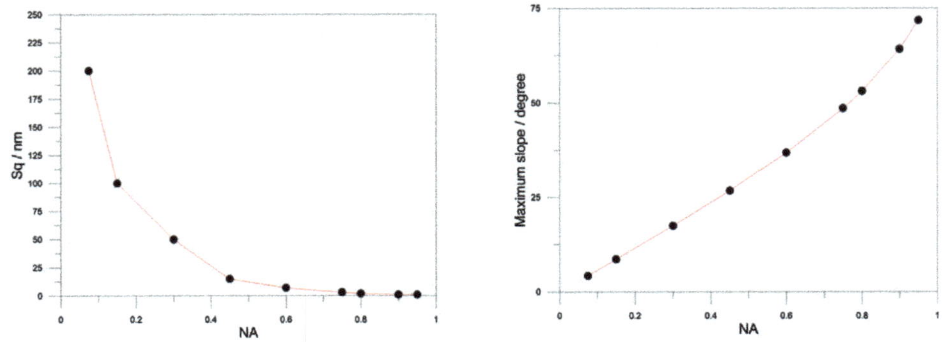

Figure 4.6. Effect of NA on the system noise and maximum measurable slope on optically smooth surfaces.

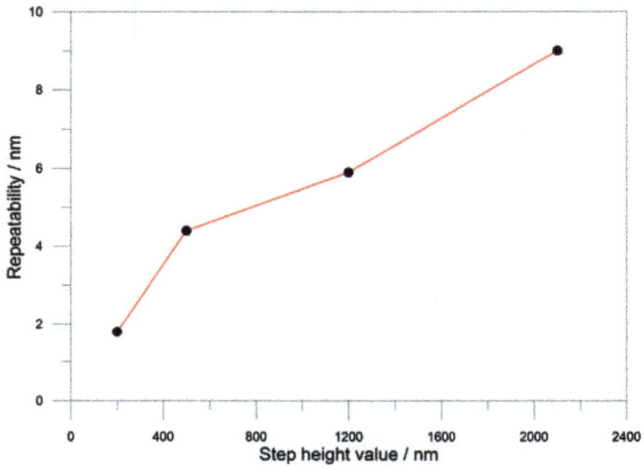

Figure 4.7. Repeatability value for step height artefacts measured with a 50× 0.8 NA objective.

smooth surfaces as a function of the NA. Instrument noise can be realised as the *Sq* value of the difference between two consecutive measurements taken on a flat mirror placed perpendicular to the optical axis. It is shown that the *Sq* value decreases with an increase in NA, reaching values close to 1 nm for NA larger than 0.8, corresponding to objectives with a magnification equal to or larger than 50×. The maximum measurable slope on optically smooth surfaces reaches a theoretical value of 72° for NA equal to 0.95, that is the maximum value achievable for objectives imaging in air (Artigas 2011). Higher NAs are possible with water and oil immersion objectives, but they are only practical in research or other specialist environments.

The vertical scanner is one of the most relevant components of an ICM, since the axial response location along the *z*-axis will be referenced to the axial position provided or positioned by the scanner. The *amplification coefficient* (α_z) as discussed before and *linearity* (l_z) (see Leach *et al* 2015 for definitions) are directly proportional to the performance of the scanner. The repeatability and reproducibility of a measurement are affected by the metrological characteristics of the system and the scanner performance. Figure 4.7 shows the repeatability value for several step height artefacts measured with a 50× 0.8 NA objective ICM.

A typical vertical scanner is a piezoelectric actuated flexure stage with a capacitive or piezoresistive position sensor. This type of scanner can provide 1 nm closed-loop positioning accuracy and, when used in combination with an ICM, the system accuracy and repeatability can also be close to 1 nm. The drawback is its limited travel range, which is typically between 0.1 mm to 0.4 mm, and thus only usable for highly flat surfaces and samples that only require a limited scan range. A larger range scanning is provided by linear stages actuated by servo, linear or stepper motors and with closed- and open-loop versions. Independent of the actuation technology, any stage will exhibit a periodic structure along its scanning range. This periodicity can come from the double interpolation error from an optical encoder, which is approximately 1/50e period of the optical grating, one revolution of the lead screw, one revolution of the motor shaft, any multiple of the reducer gearing of the motor if it is used and other factors (Arbide *et al* 2018). Figure 4.8 (left) shows a typical error position along the z-axis for a 150 μm scan with a stepper motor actuated system. A periodic structure every 20 μm can be seen, corresponding to one full revolution of the motor shaft. The amplitude of the position error is around 200 nm for this particular case. When measuring a simple structure, such as a step height artefact, there are only two locations of the scanner that are relevant to the accuracy: the z-position corresponding to the focus on the bottom of the step and the z-position focusing at the top. In the particular case shown in figure 4.8, a step height value of 10 μm will have a maximum error of 200 nm for those locations where the step is located at one valley or one crest of the positioning error function. The same step will have smaller accuracy error if it is scanned through a different location of the motor. Figure 4.8 (right) shows a simulation of a 1 μm step height measurement moving across the position error. The error inferred by the stage is ±30 nm. By performing a repeatability test, every scan will provide a moving window on such an error, averaging out and improving the accuracy of a measurement.

As with all instruments, ICMs are sensitive to external vibration. A typical industrial location vibration profile is shown in figure 4.9. The curves VC-C and

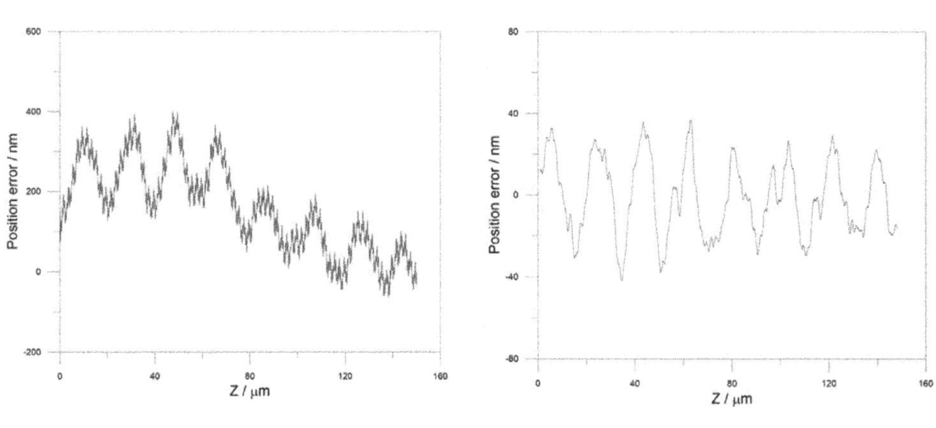

Figure 4.8. Position error of a stepper motor linear stage (left) and the inferred error on a 1 μm step height artefact (right).

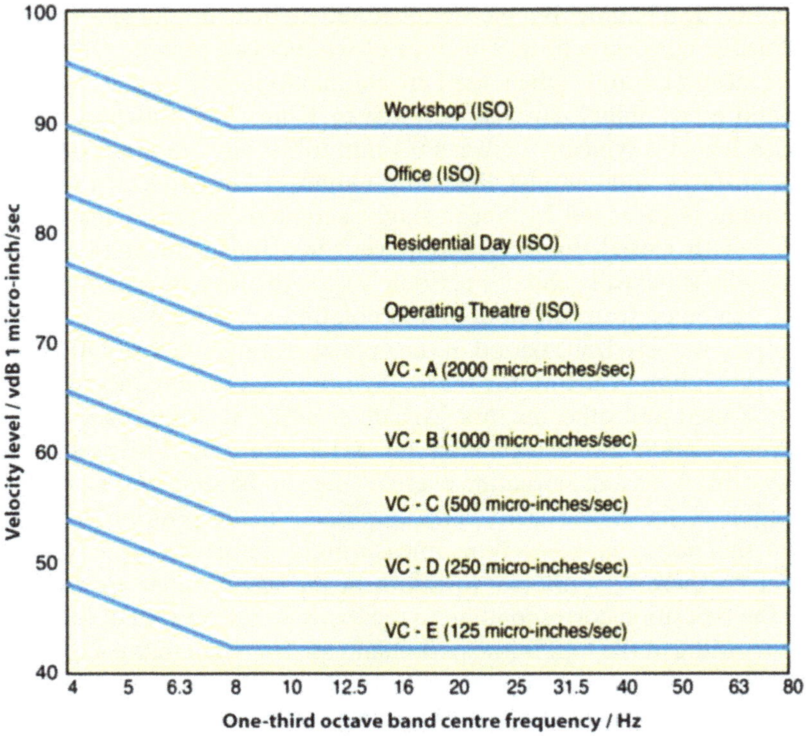

Figure 4.9. Vibration velocity with vibration frequency.

VC-E correspond to vibration environments found in many metrology laboratories and semiconductor factories, respectively.

During a vertical scan, the axial position of the scanner may not correspond to the true position on the measured part. The influence of vibration is a function of the width of the axial response and the frame rate of the confocal images. Low magnification objectives have low NA, and thus wide axial responses, so small changes in its *z*-location have a small influence. Figures 4.10 and 4.11 show a repeatability test performed with a 50× 0.8 NA objective on a 1 μm step height artefact, and a 0.9 μm *Sq* roughness artefact for both VC-C and VC-E vibration environments and a 35 fps ICM, respectively. It is clearly shown that step height structures are more sensitive due to the presence of two isolated measuring locations and to those frequencies close to the frame rate of the ICM. In contrast, the rough sample averages many *z*-positions and is less affected by vibration. The red and black curves on figures 4.10 and 4.11 correspond to VC-C and VC-E vibration profiles.

4.2.2.3 Flatness error and residual flatness error

(Adapted from Martinez 2018.)

When measuring a flat sample, such as a high-quality mirror, all ICMs show a complex topography of low spatial frequencies instead of a uniform flat result. ISO

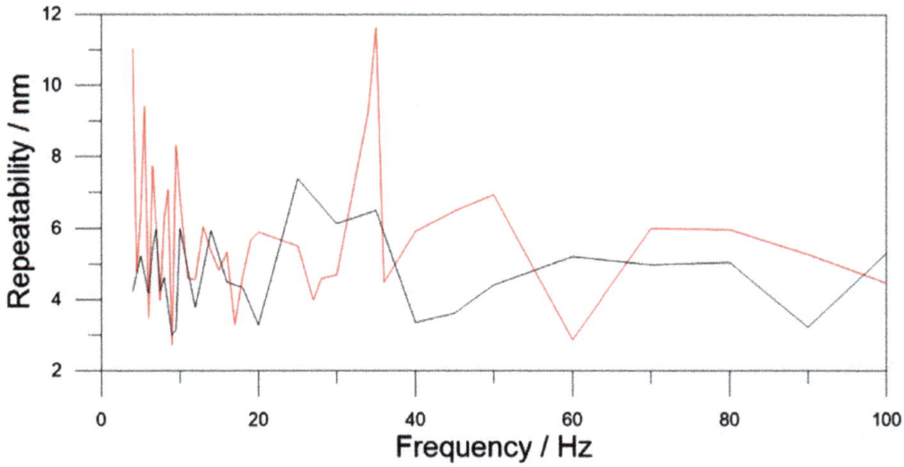

Figure 4.10. Repeatability of a 1 μm step height artefact with a 50× 0.8 NA objective for several vibration frequencies. VC-C (red) and VC-E (black) vibration profiles.

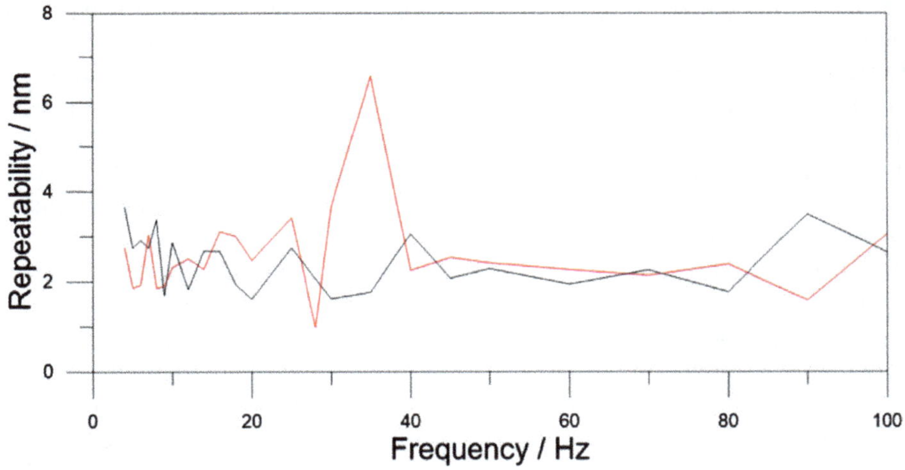

Figure 4.11. Repeatability of a 0.9 μm *Sq* roughness artefact with a 50× 0.8 NA objective for several vibration frequencies. VC-C (red) and VC-E (black) vibration profiles.

25178 part 607 (ISO 2019) states that a $\lambda/10$ flat surface with less than 0.5 nm *Sa* surface texture should be measured and the resulting surface topography used as a reference of the flatness error to be subsequently subtracted in following measurements (Giusca *et al* 2012). Figure 4.12 shows the result of measuring a flat optical mirror for three different objectives (20×, 0.45 NA, 50×, 0.8 NA and 100×, 0.9 NA) before and after the subtraction process.

The subtraction process is valid only for those surfaces that have small slopes. Nevertheless, when the object imaged through the microscope is tilted, the effective NA changes along the pupil of the microscope objective and the field curvature changes (Bermudez *et al* 2018). This changes the flatness error, leaving an additional

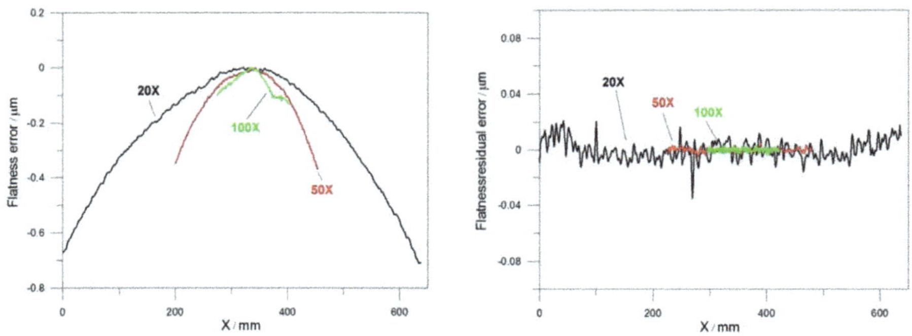

Figure 4.12. Flatness error (left) and adjustment (right) for 20×, 50× and 100× objective magnification.

error, residual flatness error, the amplitude of which is proportional to the local slope of the surface.

When measuring cylindrical surfaces, an ICM cannot measure the topography along a full revolution (Matilla *et al* 2017). The sample has to be fixed onto a rotational stage, and several topographies have to be acquired and stitched at different rotation angles. With this method, the residual flatness error is particularly detrimental, as stitching will not be accurate due to curvature mismatches.

Table 4.1 shows the residual flatness error for two different confocal arrangements: LSCM and PAM. The figure shows a flat mirror with increasing tilt from 0° to 10° with a 10×, 0.3 NA objective. At 0° tilt, the flatness error matches the expected field curvature of such an optical system with an object perpendicular to the optical axis, whereas with a finite tilt, the field curvature changes and loses its symmetry of revolution (table 4.1).

One method to adjust the residual flatness error is proposed elsewhere (Béguelin *et al* 2019). The method follows a similar concept to the correction using a flat surface but uses characterisation data for the flatness error as a function of the slope and the distance to the optical axis. The steps to realise the method are the following:

1. Measure a set of topographies with variable tilt along the x-direction and subtract the profile starting from the optical axis to the edge of the measurement.
2. Remove the tilt and construct an areal surface with the distance to the optical centre in the x-axis, the tilt angle on the y-axis and the flatness error on the z-axis. Find a mathematical function that best fits the previous surface.
3. With a new measurement, calculate the local slope and the distance to the optical centre for all the pixels and subsequently subtract the error.

The first step to characterise the residual flatness error is to measure a flat surface with different tilt angles in the x-direction, including positive and negative angles. The central x-profile is extracted and, after removing the tilted plane, some symmetry is observed—the residual profile at certain angles is symmetric with respect to the centre, compared to the same residual profile with the negative angle, as shown in figure 4.13.

Table 4.1. Flatness error of a mirror with different tilt angles for each type of ICM.

Tilt	LSCM	PAM	Scale
0°	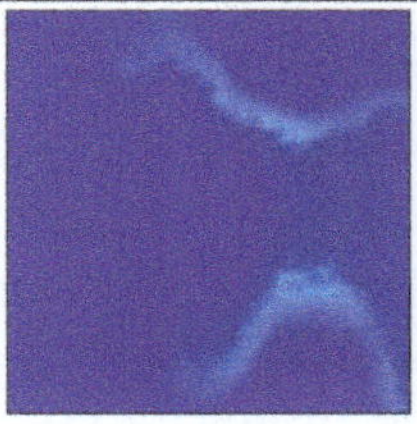		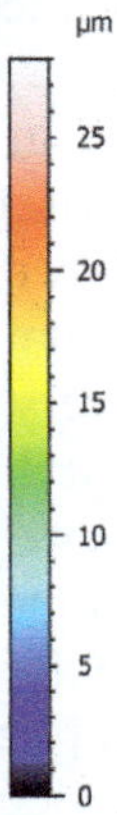
5°			
10°			

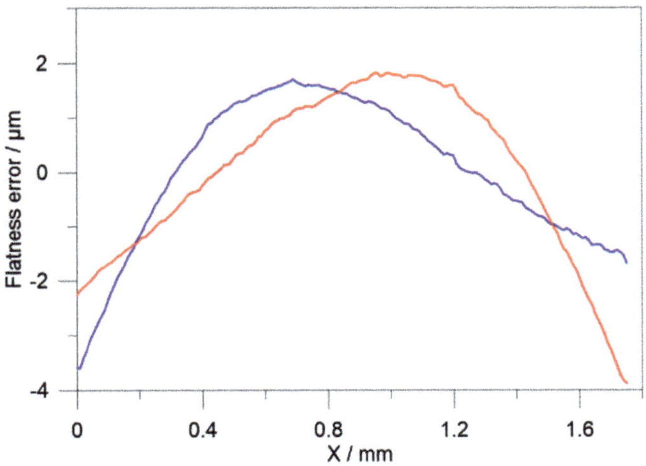

Figure 4.13. Residual flatness error for opposite slopes taken with a 10× objective on a PAM. Blue curve for 4.5°, red curve for −4.5°. Reprinted with permission from Bermudez *et al* (2018), copyright SPIE.

The symmetry shown in figure 4.13 is expected, since the flatness error is caused by field curvature aberration, which has symmetry of revolution along the axis of the optical system. On a tilted profile, the left side from its centre (the optical axis will be almost on the centre) has a certain tilt angle pointing to the centre, while the right side has the inverse tilt angle.

The same flatness error behaviour is observed in the *y*-direction: a *y*-profile crossing the optical axis of a tilted surface in the *y*-direction is equivalent to an *x*-profile of a tilted surface in the *x*-direction, for the same tilt angle. Thus, characterising one tilt direction allows extrapolation to the full flatness error, by adjusting the tilt in the radial direction and the distance to the optical centre.

The characterisation of the residual flatness error at different slope angles is made by fitting the error as a function of the distance to the optical centre and the tilt angle. A tilted flat surface in the *x*-direction is measured from 0° to 10° for a 10× objective every 1°. For each topography, the profile crossing the optical axis and parallel to the tilt direction is extracted. The first and last point of the profile are used to measure its slope and this slope is removed. Only positive slopes are measured due to the radial symmetry: the right half of the profile of a positive slope is equivalent to the left half profile if measured with inverse tilt angle and mirroring the profile with respect to the optical centre. To simplify the process, the profile is split into two halves and the optical centre is set as the origin. The right half will correspond to the positive slope, after mirroring in the *x*-axis, and the left half will correspond directly to the negative slope.

Figure 4.14 shows the areal surface topography reconstructed using the flatness characterisation method. This topography extracts the residual flatness error as a function of the radial slope and the distance to the optical centre.

An optimum mathematical function that fits the data in figure 4.14 is an eighth order Chebyshev polynomial (Bermudez *et al* 2018). Other mathematical functions,

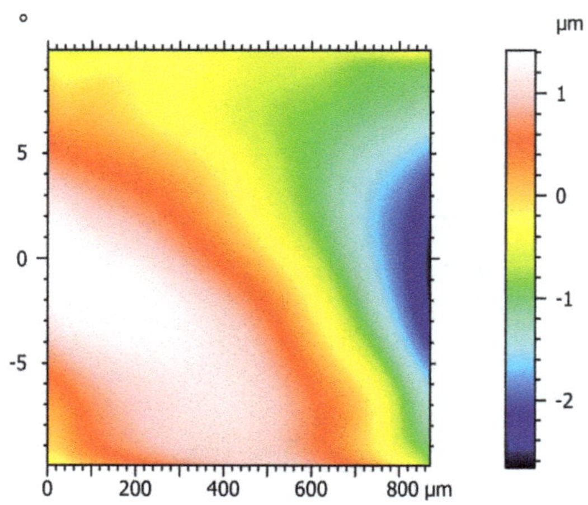

Figure 4.14. Residual flatness error surface reconstructed from the profiles of all angles measured. The *x*-axis is the distance to the optical centre (in micrometres), the *y*-axis corresponds to the tilt angle (degrees) and the *z*-axis is the residual flatness error (in micrometres). Reprinted with permission from Bermudez *et al* (2018), copyright SPIE.

such as higher order polynomials, can be more efficient from a computational point of view, but they provide less accuracy, leaving residual measurement artefacts. The Chebyshev polynomial presents a drawback in that it is not able to correct those points on the topography that have larger distances to the optical axis than the distances that are characterised. Such points are at the edges of the field of view and are due to the mathematical limitations of the Chebyshev function, which is given by

$$f(x, \alpha) = \sum_{n}\sum_{m} c_{n,m} T_n(x) T_m(\alpha), \tag{4.1}$$

where $c_{n,m}$ is the coefficient for each term, and $T_n(x) = \cos(n \cdot \operatorname{arcos}(x))$, where the x values are normalised to the interval $[-1, 1]$ from the original data, and n and m are the number of coefficients, typically eight. Because of the presence of the arccosine function, the error calculation is limited to the region where the radial distance is equal to or smaller than the characterised profile maximum distance. This causes those points of the topography with larger distances not to have correction values.

Table 4.2 shows the results of a flat mirror measured with different tilt angles, and then the dominant plane is removed from the surface to obtain the residual flatness error, and it is compared to the different calibration methods. The figures in table 4.2 show how the original residual flatness error is deformed when a tilt is applied, but does not add amplitude to the maximum error. ISO 25178 part 607 states that there

Table 4.2. Residual flatness error of a flat mirror with different tilt slopes and each method for the aberration correction.

	Levelled original measurement	Residual flatness error (ISO 25178-607)	Residual flatness error (optimised method)	Scale
0°	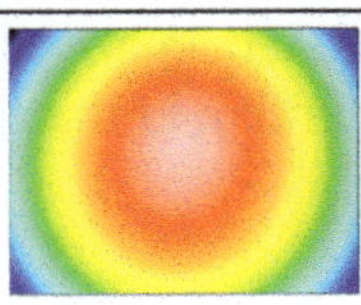		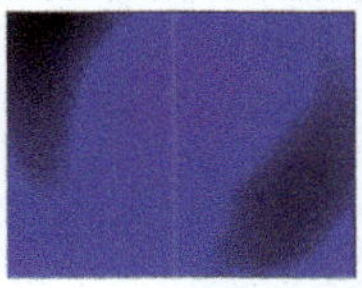	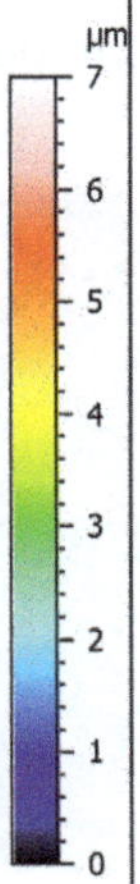
5°	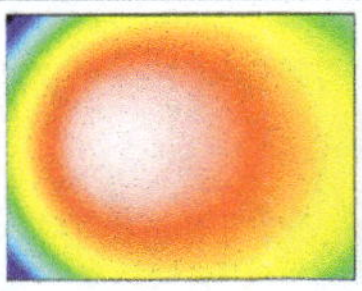	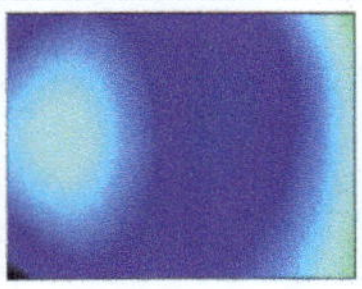		
10°	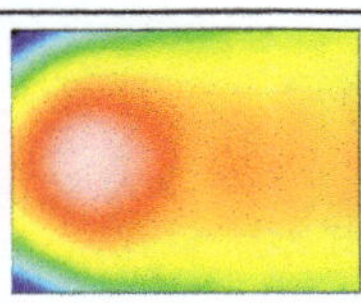	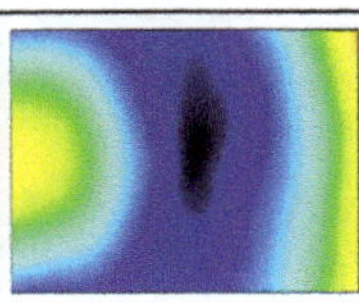		

should be Gaussian noise as the error when the surface is totally flat and levelled, but form appears when the tilt angle increases. The calibration method for the residual flatness error produces a low amplitude error with flat and levelled surfaces, and does not increase significantly in amplitude with tilt.

4.3 Structured illumination microscopy

Structured illumination microscopy (SIM) (Heintzmann 2006) is another optical sectioning method that provides similar images to those from ICM. The optical sectioning capability of SIM is similar to that of ICM when it is applied to 3D measurement of surfaces. In SIM, a periodic sinusoidal pattern is placed on the field diaphragm position of the microscope, which is imaged onto the surface. Those parts of the surface within the focal region of the objective preserve the high contrast of the original pattern, while decreasing the contrast for those areas out of focus. By phase shifting the original pattern at three different phases equally spaced over 2π, it is possible to recover the contrast modulation of the pattern with a simple calculation. Although SIM is in principle simpler than ICM, the associated pattern shifting must be precise to avoid image harmonics (Dan *et al* 2014) and thus it is not so different from the scanning complexity with other confocal configurations. The main advantage of SIM is its scanning speed with modern high-speed cameras. For every image pixel, the intensity can be approximated as

$$I_{ij} = A_{ij} + B_{ij}\cos\left(\frac{2\pi}{p}x_{ij} + \phi_{ij}\right), \tag{4.2}$$

where A_{ij} is the non-structured image (DC component), B_{ij} is the amplitude contrast of the structuring pattern multiplied by the intensity distribution of the surface, p is the period of the grating and Φ_{ij} is its phase. Figure 4.15 shows three images of a sinusoidal pattern projected onto a rough surface with a 20×0.45 NA objective. The resulting computed image is shown on the right and is similar to that from ICM.

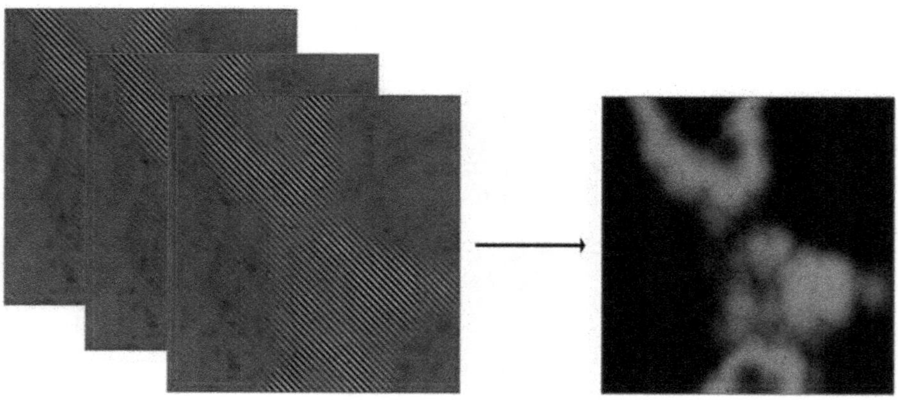

Figure 4.15. Slit images on an SIM (left) and the computed optical section (right). Reprinted with permission from Martinez *et al* 2020, copyright SPIE.

The optically sectioned image is computed from equation (4.2) by introducing a known phase shift on the projected sinusoid and acquiring a series of images. It is straightforward to show that by introducing a shift of 1/3 of the period (equally to $2\pi/3$ in phase) three times and acquiring the corresponding three images I_1, I_2 and I_3, the computed section is given by

$$B_{ij} \sim \sqrt{(I_1 - I_2)^2 + (I_1 - I_3)^2 + (I_2 - I_3)^2} \tag{4.3}$$

and the DC components of the image are

$$A_{ij} \sim I_1 + I_2 + I_3. \tag{4.4}$$

In order to introduce the shift of the sinusoid, several methods have been proposed. Most of the methods use high-precision mechanical stages or tilting optics actuating onto a glass substrate. Non-moving parts using projection microdisplays are also used, known as programmable array microscopes.

To avoid phase shifting the illumination pattern mechanically, Wicker *et al* (2010) proposed a single-shot optical sectioning method by splitting the three phases of the original pattern on three different polarisations and using three cameras with properly aligned polarisers.

Chromatic confocal microscopy is a very well-known technique for single point height measurement (Blateyron 2011). Zint *et al* (2019) developed a system using a set of pinholes equally spaced on a matrix and imaged onto to the surface with a hyperchromatic objective. The pinholes are imaged with a spectral separation unit extending the spectrum for every point along a camera's set of pixels. The design is single-shot and is capable of achieving hundreds of 3D frames/second, although the measured region contains only 1000 points and the numerical aperture is low.

A simpler implementation of SIM is described by Schwertner (2011). A sinusoidal pattern is illuminated in transmission and reflection, and two images are recorded. The fact that the sinusoidal pattern is shifted by π makes it impossible *a priori* to recover the optical section with a simple calculation. It requires a calibration process to calculate the phase of the pattern for every pixel and the calculation of the gradient along the pattern direction. Other non-moving techniques involve the use of the Hilbert transform. Krzysztof *et al* (2014) used two images, a sinusoidal pattern and a bright field arrangement to compute a $\pi/2$ shift of the original pattern using the Hilbert–Huang transform. Hoffman *et al* (2018) used the same idea to avoid scattering problems for *in vivo* imaging, where contrast of the pattern through scattering media decreases, and the phase shift is no longer trustworthy. Hoffman *et al* used a single image and recovered the bright field image by blocking the main frequency of the pattern in the Fourier space.

From equation (4.2) it is possible to isolate the non-structured DC components of the bright field image, thus

$$I'_{ij} = I_{ij} - A_{ij} = B_{ij} \cos\left(\frac{2\pi}{p} x_{ij} + \phi_{ij}\right). \tag{4.5}$$

The optically sectioned image is recovered by the amplitude contrast B_{ij} that is easily recovered with

$$I''_{ij} = G(I') = B_{ij} \times \sin\left(\frac{2\pi}{p}x_{ij} + \phi_{ij}\right), \tag{4.6}$$

where G is any operator that shifts the frequency components of the image by $\pi/2$ and thus

$$B_{ij} = \sqrt{(I'_{ij})^2 + (I''_{ij})^2}. \tag{4.7}$$

There are several approaches for the G operator. The most straightforward approach is to consider the image as a series of profiles consisting of each image column, when the sinusoid is projected in the vertical direction, or each image row if the sinusoid is horizontal. It is possible to apply to each profile the Hilbert transform, which causes an exact phase shift of $\pi/2$ to any one of them. The shifted image is then recovered by simply substituting each column or row by the shifted profile. This approach does not consider the phase gradient of the image, inducing some errors in samples that have a specific direction. Additionally, it is not a computationally efficient approach, since it needs as many fast Fourier transforms (FFTs) as there are image columns/rows, instead of just performing a bidimensional FFT.

The Hilbert–Huang transform is an extension of the Hilbert transform that decomposes the original signal into various intrinsic mode functions. This decomposition allows the removal of the high oscillation mode, which is usually associated with noise, but has the inconvenience of being computationally expensive (Xing 2015).

The main limitation of the Hilbert transform applied to images is the extension of the sign function into a 2D function. The spiral phase quadrature transform (Larkin 2001), shown in figure 4.16, solves the problem of the directional discontinuity of the sign function using a pure spiral phase function in the Fourier space, defined as

Figure 4.16. Pure spiral phase function in the spatial frequency domain. Real part (left) and imaginary part (right). The scale is −1 to 1. Reprinted with permission from Martinez *et al* 2020, copyright SPIE.

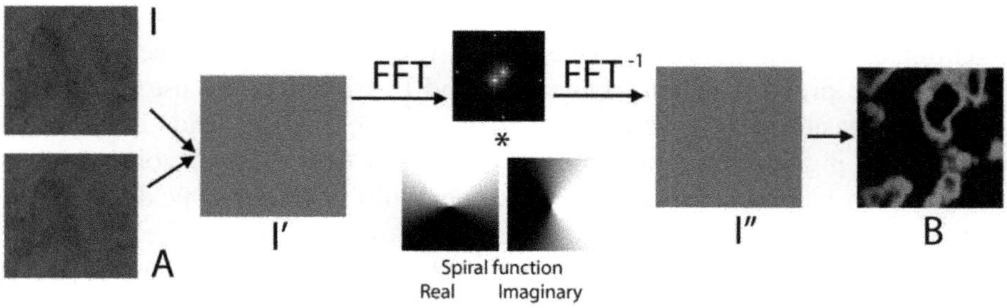

Figure 4.17. Diagram of the two-image method for optical sectioning. Reprinted with permission from Martinez *et al* 2020, copyright SPIE.

$$S(u, v) = \frac{u + iv}{\sqrt{u^2 + v^2}}, \tag{4.8}$$

where u and v are spatial frequencies.

The optically sectioned image is recovered by acquiring two images: the structured illuminated image and a bright-field image, which represents the DC component at each pixel. The bright-field image is subtracted from the structured illumination image to obtain I', the spiral phase quadrature transform is applied to obtain I'', and finally equation (4.7) is applied to recover the optically sectioned image B. Figure 4.17 shows the process to digitally compute an optical section.

This method requires a projection of the sinusoid pattern onto the sample and uniformly illuminates the sample to recover the bright-field image. This causes more complexity on the system set-up or obligates the presence of a microdisplay component, which can significantly increase microscope costs.

4.4 Simultaneous confocal and focus variation in a single acquisition scan

(Reproduced in part from Matilla *et al* (2016).)

It is difficult to measure smooth surfaces with focus variation, since no texture is present on the surface and no focus position can be retrieved (but see recent advances in chapter 3). Surfaces that at a given wavelength and NA appear optically smooth (and are thus not suitable for conventional focus variation), may appear as optically rough when decreasing the wavelength or magnification. This is the reason why focus variation is most typically suitable with low magnification, since most of the surface then appears as optically rough.

There is a growing demand for data fusion from the macro-scale to the micro-/nano-scale. This is the case where large parts are manufactured within tight tolerances or with small-scale features. Diamond turned optical surfaces with diffractive patterns are a typical case where the lens may be several millimetres in diameter while having topographical features of the order of a few micrometres wide and less than 1 μm in height. Full measurement requires gigapixel information, which can only be carried out by the stitching of sampled fields. Another typical case is large metallic parts with micro-machined surfaces, which cannot be sampled with conventional contact coordinate measuring machines (CMMs) because of the dilation of the probe with the topography and this requires the use of an optical coordinate measuring systems (Leach 2020) to sample small fields of view while preserving the high accuracy of the measured field within a larger volume.

Data fusion has been carried out since the 1960s in different fields and can be primarily classified into three levels: decision, feature and signal level (Wang *et al* 2015). Decision data fusion is when a set of parameters describing the surface characteristics are extracted from the topographical data, such as texture or step height data, from measurements taken from different sensors and at the same or different scales. The data collected from the sensors provides the necessary information to take a decision, such as sorting the parts into different quality levels or binning parts within different manufacturing plants. Feature extraction data fusion is of a lower level, where geometrical structures are extracted from sets of topographical measures with different scales. Signal level data fusion relies on the topographical correlation of data from the same or different sensors, providing point cloud data as a result. When dealing with separated sensors at different scales, topographical data has to be resampled (Ramasamy *et al* 2013) and registered before fusing the data. The most common signal level fusion is that from the same sensor with the same measuring technology. In this case, there are several uses of the data:

- Data fusion from different fields at the same scale to cover larger measured areas. This is known as stitching or sub-aperture stitching, common in many commercial 3D scanning instruments. Stitching is carried out by translating the sample being measured to different positions in either a two-axis *xy* stage or a five-axis scanner, with an overlapping area between fields to aid post-processing during data registration.
- Fusion of topographical data acquired with different magnification objectives to provide form and texture in a single result (Ramasamy *et al* 2013).
- Data fusion within the same field of view and same magnification, but with different scanning parameters. This is the typical case where a sample has high reflectivity and low reflectivity regions, requiring different illumination levels to deal with the dynamic range of the imaging camera. Several strategies have been adopted such as using two vertical scans at different light levels (Fay *et al* 2014), using HDR cameras, or coating the sample with fluorescence polymers to increase the scattered signal at high slope regions (Liu *et al* 2016).
- Data fusion over time on the same field with the same scanning parameters to deal with instrument noise and external disturbances. Data is averaged providing lower noise and higher repeatability.

Topographical fusion of data coming from different sensors on the same instrument that can be operated simultaneously is also possible. The benefit of acquiring the data on the same instrument along the same scan is that most of the data registration is avoided, providing a result with higher linearity and accuracy. When fusing the signal data from two sensors at least five different characteristics of the sensor itself have to be taken into account before registration:

1. *Levelling.* This is a simple step if the surface under inspection is relatively flat, but difficult if step-like structures are present. Levelling has to be carried on the two topographies by extracting features and correlating them. On rough random surfaces, accurate levelling is problematic.
2. *Calibration of the lateral scale.* Similar to levelling, if structures are present on the surface, they can be extracted and registered, but accurate calibration of the optical magnification and field distortion is necessary before data fusion.
3. *Data post-processing.* Outlier removal, missing points, filling non-measured regions, and smoothing are different post-processing steps carried out for different measuring technologies. Data correlation, registration and interpolation are often necessary.
4. *Linear amplification coefficient.* If the data are coming from two different technologies, calibration of the linear amplification should be carried out with care. This is typically achieved with a set of traceable step height artefacts by measuring them and calibrating the slope of the linear scan to minimise the error (Leach *et al* 2015a). This process has to be taken into account within the same instrument with different sensors: an ICM drives the scanner step-by-step while coherence scanning interferometry (CSI) or focus variation move the scanner at continuous speeds; the scanner linear amplification coefficient could be different in these cases.
5. *Stage linearity.* Even with accurate linear amplification calibration, non-linearities of the scanner are superimposed onto the topographical data. Different pixels at different measured heights will have different errors, which will confuse the data registration and thus provide lower accuracy on the fused topography than for the case of a single measurement.

The optimum instrument to cater for measurements on as many different kinds of surfaces as possible while avoiding the aforementioned problems will be the one that has the capability to perform measurements with many of the imaging technologies during the same scan. Nevertheless, there are some surfaces where none of the three aforementioned technologies (CSI, ICM and focus variation) yield ideal results. The combination of data from two of the three technologies could in principle provide improved results. ICM and focus variation have the optical sectioning property in common, with similar depth-of-focus characteristics. In general, the main difference between them is that ICM deals better with smooth surfaces, while focus variation is more optimum with rough surfaces.

Many commercial ICMs are able to acquire a bright-field image with a separate dedicated imaging channel. On these instruments, the scanner is driven in a step-by-step manner, providing two series of images, confocal and bright field, that are used to fuse data information from confocal and focus variation. By doing this, levelling is the same for both sets, objective magnification and field distortion are the same, the linear amplification coefficient is the same, and non-linearity of the vertical scanner is superimposed on the two data sets with the same pattern. This process avoids correlation, data registration and interpolation, providing as a result with higher accuracy than many other data fusion techniques.

The focus variation technique is similar to confocal microscopy in that the images have optical sectioning properties. The main difference is that confocal images have real optical sectioning for each pixel, while the sectioning ability of focus variation relies on the texture of the surface, the numerical aperture of the objective, the wavelength, the focus algorithm, and the algorithm to fill poor signal regions. Typical focus operators are (Helmli *et al* 2001) Laplacian, sum of modified Laplacian (SML) and gradient grey level (see also chapter 3). Each different operator gives different results depending on the texture of the surface and the numerical aperture of the objective.

The fusion of the data coming from ICM and focus variation is a straightforward process. With this method, the two series of images result in two measured topographies. The most accurate of the two will be that from the ICM, but it will typically have a greater number of non-measured points when a safety threshold is used. If the signal-to-noise ratio (SNR) is low, the resulting measured point could be a spike or non-measured point. The focus variation topography is less accurate, but due to its algorithmic nature, it will provide topographical data on high slope and rough regions, despite typically larger instrument noise compared to ICM. Topographical fusion is achieved by identifying the non-measured points on the confocal topography and creating a mask that is applied to the focus variation topography. This masked result is smoothed and copied to the confocal topography. Figure 4.18 shows the result of this method on a micro-machined surface. The

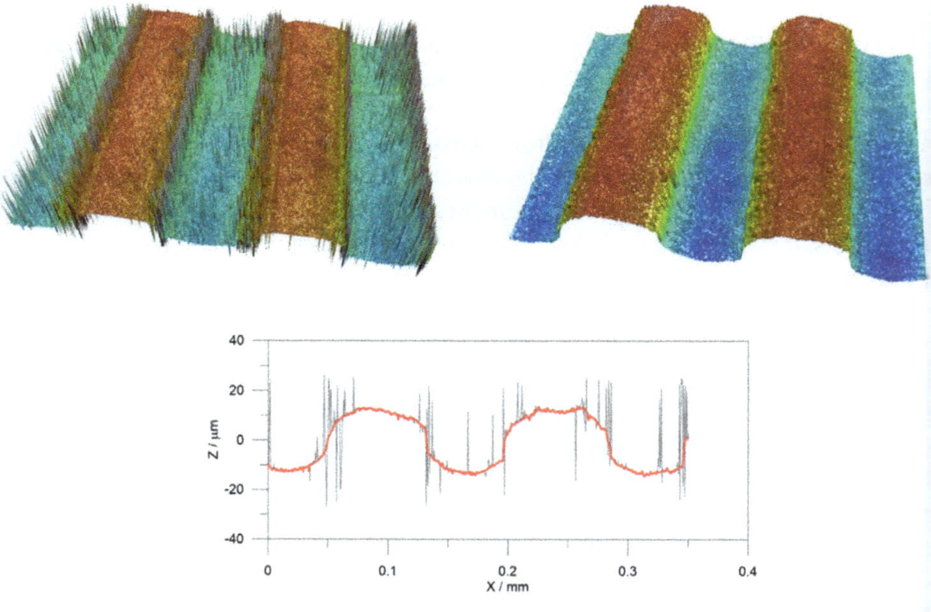

Figure 4.18. 3D measurement of a micro-machined surface with confocal (above, left) and after topographical fusion with focus variation (above, right). Shown below are cross-sectional profiles of the confocal data (light grey) and of the fused data (red line). Reprinted with permission from Artigas *et al* (1999), copyright SPIE.

topographies (left/right) are the confocal and after topography fusion, respectively, while the profiles shown below are of the raw confocal (in light grey) and the fused data (red line). For better understanding of the loss in confocal mode on low signal regions, a zero threshold and no post-processing of the confocal topography is shown. The spikes shown in figure 4.18 belong to non-measured points when a safety threshold is used.

Another image fusion technique is a plane-by-plane approach. Unlike the topographical method, this lower level approach fuses information by creating new averaged plane images from which the 3D result is computed. The main benefit of this method is the addition of information in the stack of images, in areas that are non-measured, in one single technique. However, the fusion smooths other areas that are already well defined.

The optical sectioning in confocal and focus variation is proportional to the wavelength and inversely proportional to the square root of the numerical aperture. The z-resolution value of the optical sectioning could slightly vary depending on the sample's texture, the objective used and focus algorithm applied. The resulting ICM and focus variation images at each plane share many similarities, with more depth discrimination in the confocal case.

In this confocal and focus variation fusion approach, the mean value of each image pair is computed, and then the focus variation image is offset to match the signal of the confocal image. By doing this, the confocal series retains the original signal, while the focus variation is dynamically adjusted. Figure 4.19 shows how the image fusion is able to enhance the information in areas in which the confocal image is weak. This plane-by-plane approach results in a third, fused series of images, from which the 3D result is computed.

A third method uses the information of the axial response on a pixel-by-pixel approach. Similar to the topographical and the image fusion technique, this approach combines the information of confocal and focus variation, gaining improved results over any single technique. However, in this case, the method is able to dynamically select what is the best axial response for each pixel: confocal, focus variation or fused. The noise is reduced for those pixels where both techniques have a weak peak by averaging both axial responses and generating a defined peak. Figure 4.20 shows two results for fusing the axial response signal. The image on the left shows a region of the surface where the confocal signal is low and focus variation is high. The image on the right shows a region of the surface where both signals have a poor SNR. The fused information improves the SNR in this case.

Figure 4.21 shows a cross-section profile of a laser drilled copper surface along the highest slope region. The grey profile is the raw confocal result, with a noisy and spiky appearance on the slope regions, and smooth and accurate data on the flat regions. The red profile is the result of axial response fusion.

Figure 4.22 shows the result of a topography measurement on a cooper plate with a 50×0.8 NA objective. Confocal and focus variation topographies are shown in the

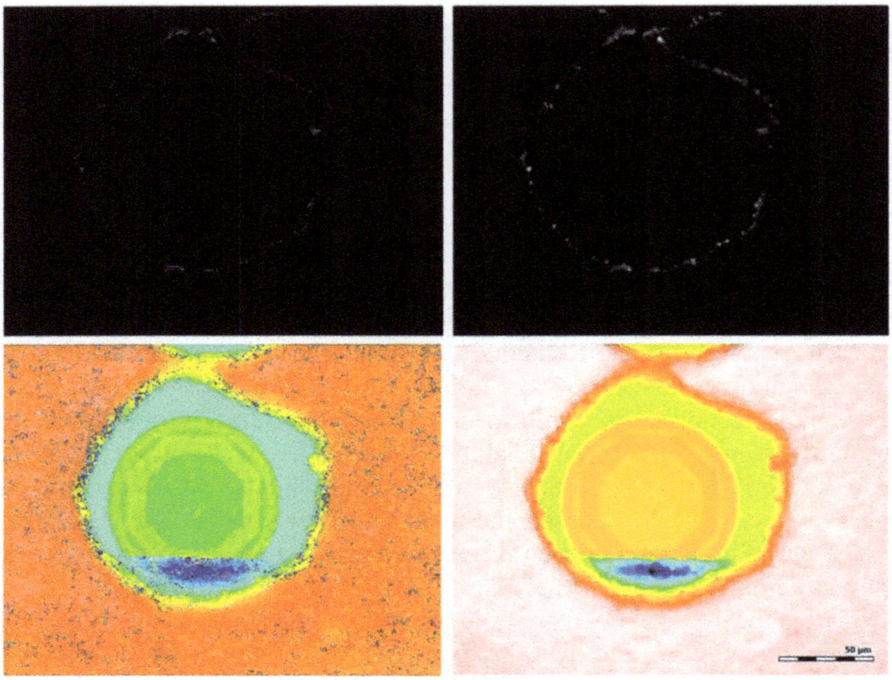

Figure 4.19. Comparison at the same plane with confocal image (above, left) and image fusion (above, right). The confocal image signal is really low and its 3D result (below, left) at this plane implies non-measured points while the image fusion enhances the signal and its 3D result is improved (below, right). Reprinted with permission from Artigas *et al* (1999), copyright SPIE.

Figure 4.20. Axial responses of confocal (blue), focus variation (green) and after fusion (red). Left: axial response where focus variation imposes over the confocal signal. Right: axial responses with similar and very low SNRs. The fused axial response is showing improved SNR. Reprinted with permission from Artigas *et al* (1999), copyright SPIE.

top image. The higher lateral resolution of the confocal result is noticeable, as is the lower the resolution of the focus variation, due to the focus operator itself. On the lower part of figure 4.22, the results from the three fusion algorithms are shown (topography fusion, image fusion and axial response fusion).

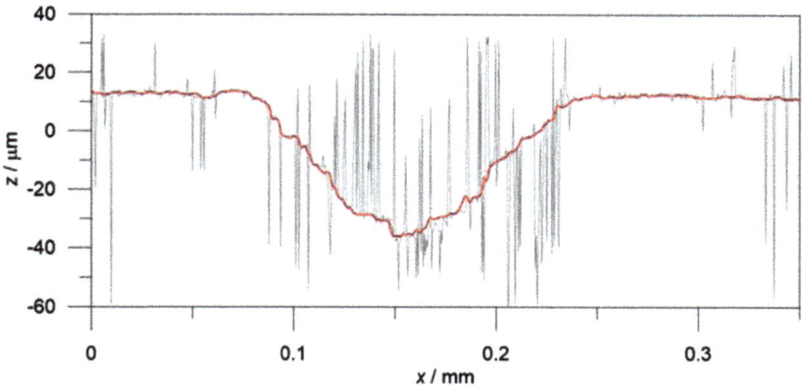

Figure 4.21. Profile across the deepest part of the topographies shown in figure 4.19. Light grey is the confocal result, while red is the profile from the image fusion technique. Reprinted with permission from Artigas *et al* (1999), copyright SPIE.

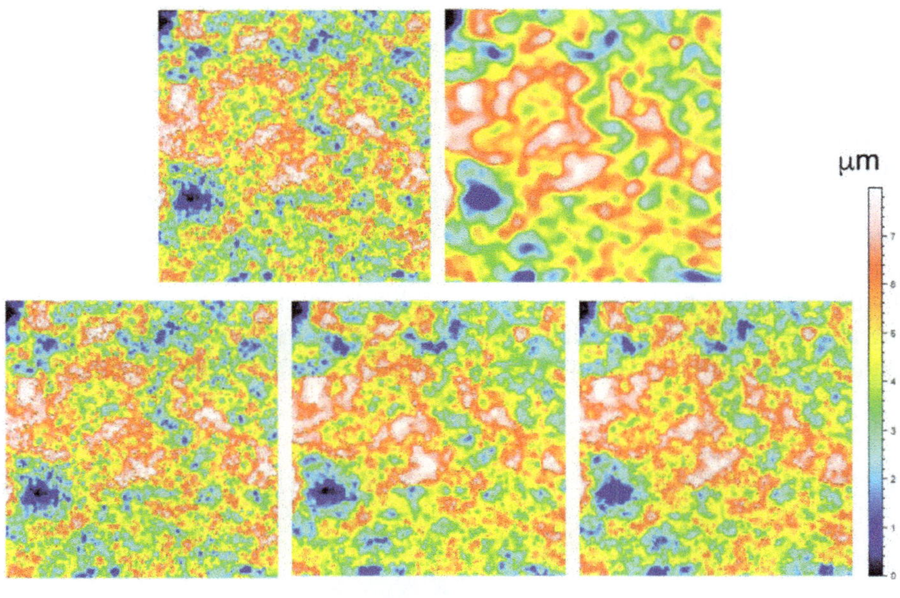

Figure 4.22. Zoomed area of figure 4.19 on the copper surface. Top left confocal, top right focus variation. Bottom from left to right: topographical fusion, image fusion and axial response fusion. Note the loss in lateral resolution of the focus variation image and a recovery of image detail on the three fusion techniques. Reprinted with permission from Artigas *et al* (1999), copyright SPIE.

4.5 Active illumination focus variation: the 'poor person's confocal microscopy'

(Reproduced in part from Bermudez *et al* (2019).)

Noguchi *et al* (1994) proposed a method to actively illuminate the surface under evaluation (to get around the need for a degree of texture with focus variation—see chapter 3 for other methods to avoid this issue). The pattern proposed was a chessboard with a period optimised for the numerical aperture, magnification and wavelength. The pattern is placed on the field diaphragm position of the microscope and uniformly illuminated. The chessboard is projected onto the surface and its reflection is imaged on the observation camera simultaneously with the image of the surface. For those regions with low levels of texture, or theoretically none, the projected pattern provides local contrast in focus, while for those heavily scattering regions, the pattern no longer disturbs the original local contrast. In figure 4.23, the bright field image of an type AIR material measure from the National Physical Laboratory (NPL) (Leach *et al* 2013, Nimishankavi *et al* 2019) is shown, which has been calibrated using a traceable stylus instrument. Unfortunately, the surface is optically smooth, and conventional focus variation cannot provide sufficiently good optical sectioned images. As shown in the centre of figure 4.23, a sum of modified Laplacian focus operator with a 5 × 5 window only gives a strong signal for those pixels with the small dark spots. The areal topography result shown on the right of the figure is too noisy for the evaluation of its surface texture parameters.

Figure 4.24 presents the same sample measured with the same conditions as those in figure 4.23, but using the technique described by Noguchi and Nayar (1994). The sample has been illuminated actively with a chessboard pattern placed on the field diaphragm of the microscope. The optically smooth surface is now showing high local contrast for those regions in focus, while the out-of-focus regions simultaneously defocus the chessboard pattern. The optical section computed with a 5 × 5 sum of modified Laplacian is shown in the centre. This optical section resembles a confocal image. The computed areal topography shown on the right has enough quality to evaluate the surface texture parameters. The obtained results compare well with the values provided by the NPL calibration certificate.

Figure 4.25 shows the result for measuring with this active illumination focus variation (AiFV) method and the conventional confocal image with the same

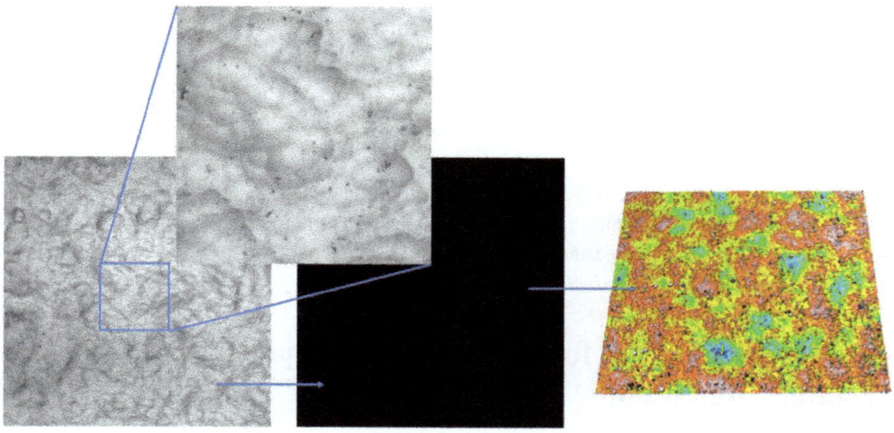

Figure 4.23. Bright field image of a type AIR material measure (left), the focus assessment of one evaluation plane with a sum of modified Laplacian operator at 5 × 5 evaluation window (centre) and the computed areal topography (right). Reprinted with permission from Bermudez *et al* (2019), copyright SPIE.

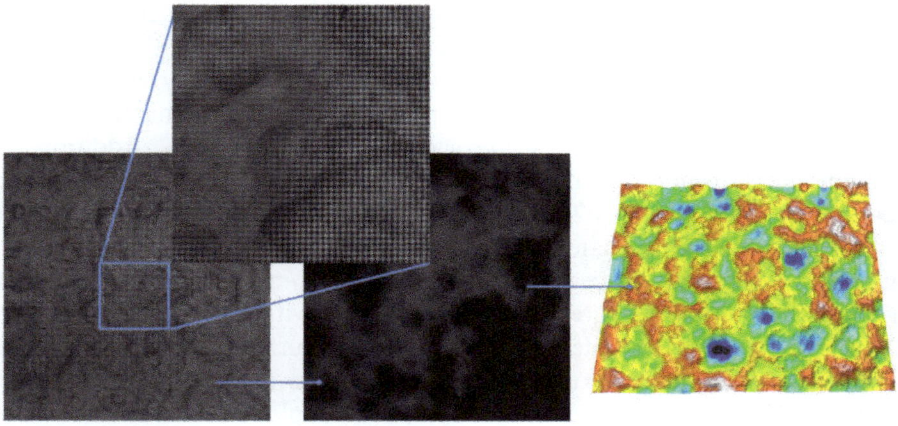

Figure 4.24. Same sample and same evaluation conditions as in figure 4.23, but with an optically superimposed artificial texture. Right: the computed areal topography. Reprinted with permission from Bermudez *et al* (2019), copyright SPIE.

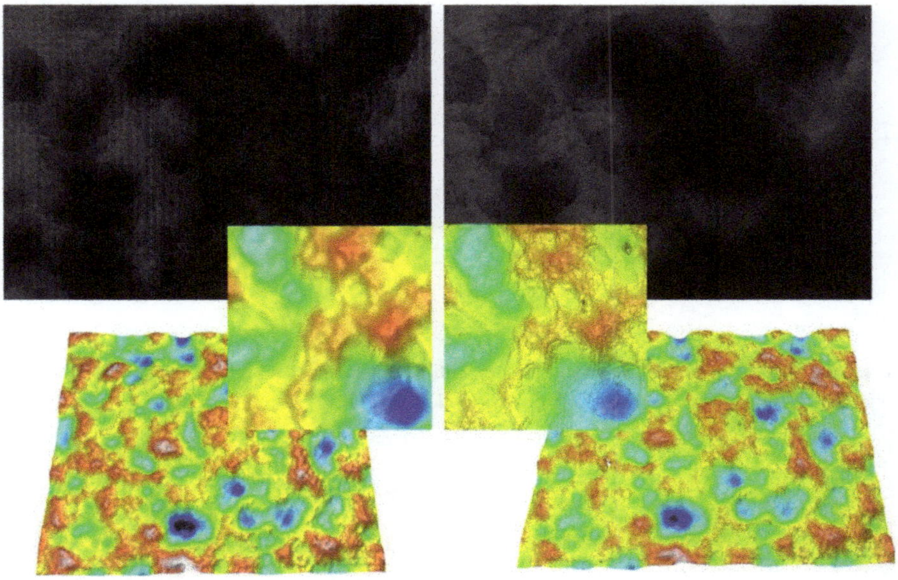

Figure 4.25. Left: optical section by applying a sum of modified Laplacian with a window of 5 × 5 pixels on an actively illuminated surface with a 20 × 0.45 NA objective, and its corresponding computed areal topography. Right: a confocal image at the same height location and its computed areal topography. Reprinted with permission from Bermudez *et al* (2019), copyright SPIE.

artefact as shown in figures 4.22 and 4.23. The image on the left shows the same result as in figure 4.25 by computing the areal topography from a set of actively illuminated images and a sum of modified Laplacian focus operator with a 5 × 5 pixel window and a 20 × 0.45 NA objective. On the right, the same surface at the same location is measured with the confocal technique. It is clearly shown that the confocal topography is able to preserve the high spatial frequency components of the surface due to its ability to provide high lateral resolution. Medium and lower spatial frequencies are well preserved with the AiFV measurement.

AiFV is capable of measuring the areal topography of optically smooth surfaces Most calibration artefacts, such as the step heights or periodic material measures, are manufactured on a flat substrate such as glass or silicon. Additionally, a chromium layer around 100 nm is typically deposited on top of the surface to increase reflectivity. Figure 4.26 shows the measured profile with confocal and an AiFV with a sinusoidal material measure manufactured on a glass substrate. Calibrated values from the certificate are Ra = 0.88 μm with Rt = 3.0 μm. Both measurements are in close agreement and consistent with the certified values.

Other material measures intended for areal optical surface measuring instruments exhibit a random type of texture with high, medium and low spatial frequency features. AiFV starts to become compromised for samples where the surface texture has a correlation length close to the evaluation window of the focus operator and the optically projected artificial texture. In order to show this limit, figure 4.27 shows a

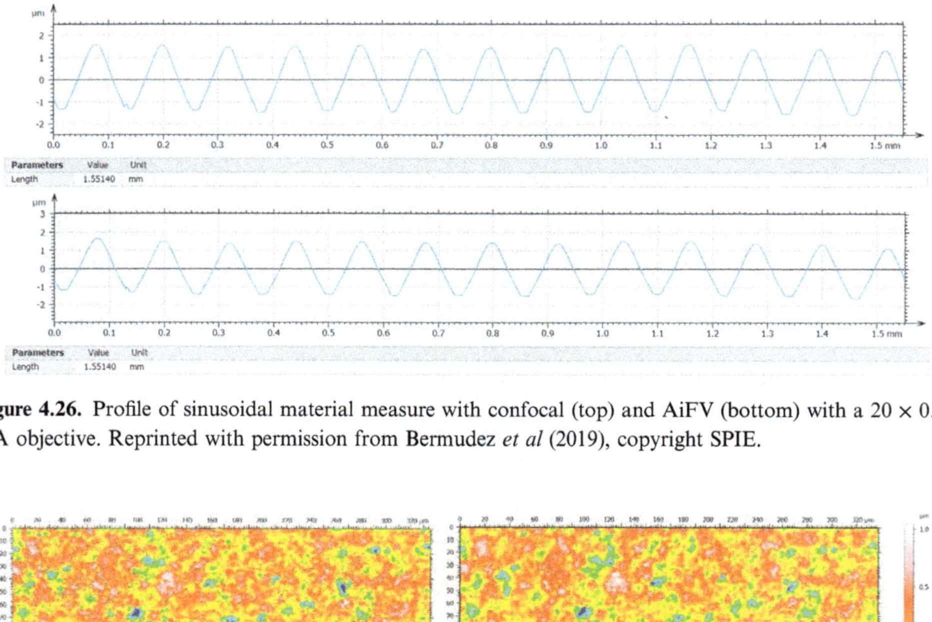

Figure 4.26. Profile of sinusoidal material measure with confocal (top) and AiFV (bottom) with a 20 × 0.45 NA objective. Reprinted with permission from Bermudez *et al* (2019), copyright SPIE.

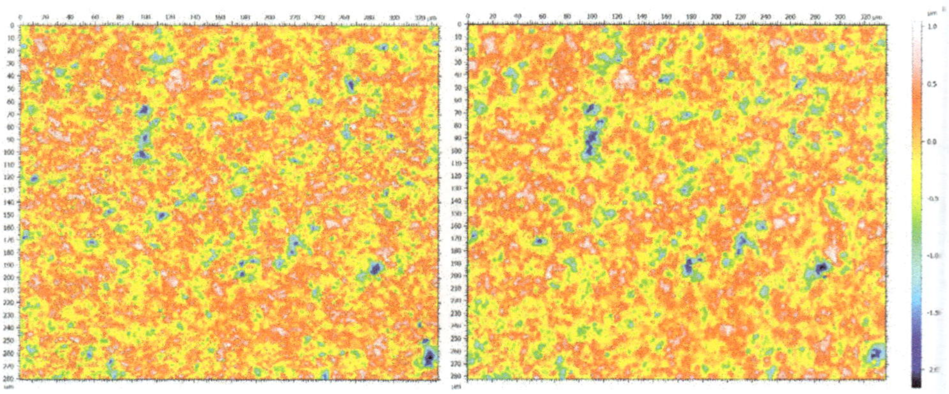

Figure 4.27. Areal topography of type ARS material measure with confocal (left) and AiFV (right). Reprinted with permission from Bermudez *et al* (2019), copyright SPIE.

type ARS material measure with Sa = 0.29 μm and a correlation length of approximately 3.5 μm, measured with a 50 × 0.8 NA objective with confocal (left) and AiFV (right). The smoothing effect shown in the figure will affect the values of the surface texture parameters, since the high spatial frequency components are close to the lateral resolution limit of the technique.

The ability to preserve high spatial frequency details on the surface is dependent on three factors: (i) the magnification and NA of the objective; (ii) the pixel size of the camera, the field lens magnification factor and the evaluation window of the focus operator; and (iii) the pixel size and pattern period of the chessboard and the magnification factor of the field lens that projects the image of the microdisplay onto the surface.

The measurement results depicted in figure 4.27 show that surface details smaller than the chessboard pattern period or window evaluation size are not resolved. The texture parameter values, both height and lateral parameters, will be affected by the system magnification. Figure 4.28 shows the result of measuring Sa and Sal (surface autocorrelation length) on the sample shown in figure 4.27 for both confocal (black dots) and AiFV (red dots). It is shown that the Sa and Sal values are consistent with a 100× or higher magnification for the confocal instrument, while for AiFV the values are less consistent. A 50× magnification still provides close results for confocal, but not for AiFV.

The lateral resolution limit of AiFV can be determined using a type ASG star-shaped material measure, following the method described elsewhere (Giusca *et al* 2013). Figure 4.29 shows the result of measuring the areal topography of such an artefact with 20× (top) and 50× (bottom) magnification objectives with confocal and AiFV. The superior performance of the confocal microscope to preserve lateral dimensions and thus higher spatial frequencies is clearly shown. Nevertheless, AiFV is also capable of performing well up to a certain spatial frequency limit. To assess this limit, the artefact was measured with a range of objectives with magnifications from 20× to 150×. Figure 4.30 shows the result of the lateral resolution limit for confocal (black dots), AiFV (red dots) and half the Rayleigh diffraction limit (blue dots). Taking into account these results, it can be stated that a 20× objective is capable of resolving down to 1 μm structures on optically smooth surfaces with the AiFV technique.

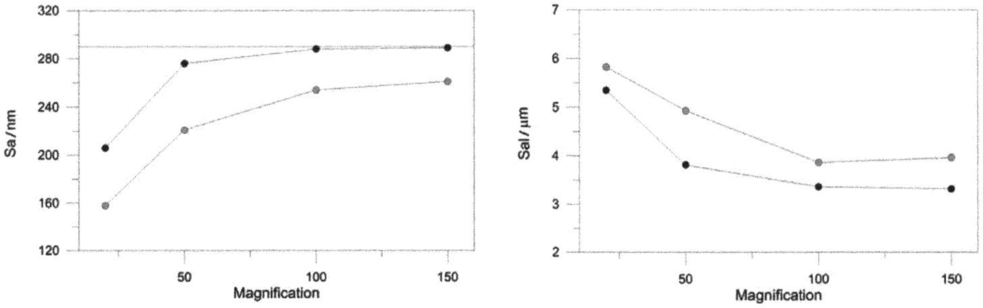

Figure 4.28. Sa and Sal for several magnifications. Black dots: confocal. Red dots: AiFV measurements. Reprinted with permission from Bermudez *et al* (2019), copyright SPIE.

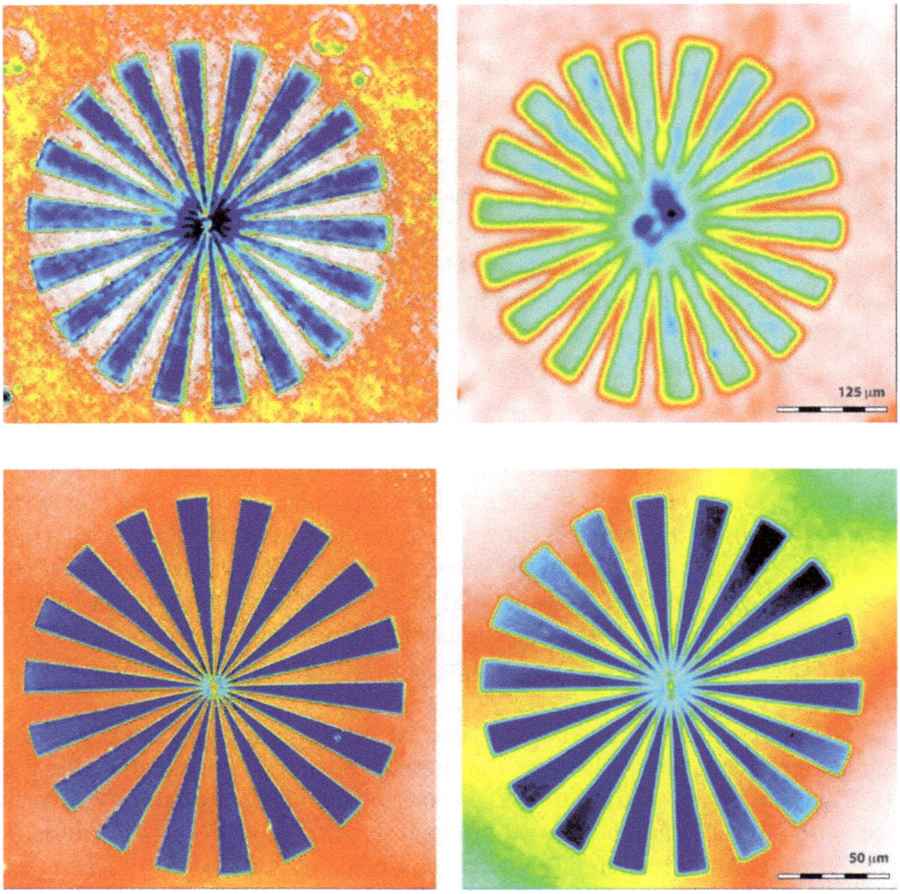

Figure 4.29. Type ASG material measure measured with a 20 × 0.45 NA (top) and 50 × 0.8 NA (bottom) and confocal (left) and AiFV (right). Reprinted with permission from Bermudez *et al* (2019), copyright SPIE.

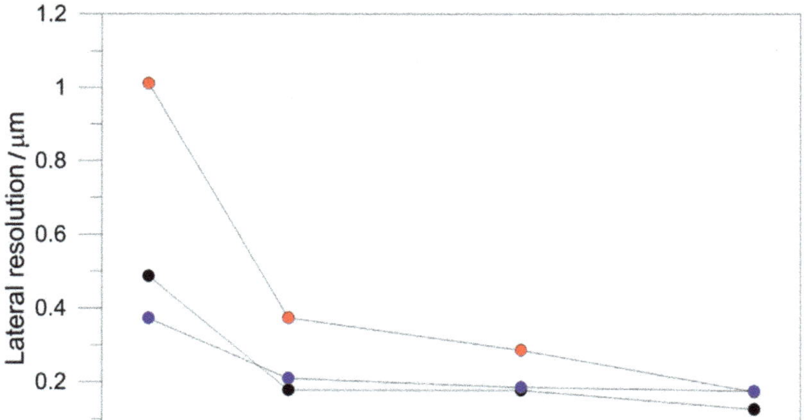

Figure 4.30. Lateral resolution derived from the Siemens star method for objectives magnification ranging from 20× to 150×. Black: confocal; red: AiFV; blue: half the diffraction limit. Reprinted with permission from Bermudez *et al* (2019), copyright SPIE.

References

Arbide A, Laguarta F and Artigas R 2018 Performance characterization of an optical profiler though the measurement of the linearity deviations of the scanner *Master's Thesis* Technical University of Catalonia

Artigas R, Laguarta F and Cadevall C 2004 Dual-technology optical sensor head for 3D surface shape measurements on the micro- and nano-scales *Proc. SPIE* **5457** 166–74

Artigas R, Laguarta F and Pinto A 1999 Three-dimensional micromeasurements on smooth and rough surfaces with a new confocal optical profiler *Proc. SPIE* **3824** 93–104

Artigas R 2011 Imaging confocal microscopy *Optical Measurement of Surface Topography* ed R K Leach (Berlin: Springer)

Béguelin J, Scharf T, Noell W and Voelkel R 2019 Correction of surface error occurring in microlenses characterization performed by optical profilers *Proc. SPIE* **11056** 110560

Bermudez C, Fegner A, Martinez P, Matilla A, Cadevall C and Artigas R 2018 Residual flatness error correction in three-dimensional imaging confocal microscopes *Proc. SPIE* **10678** 10678M

Bermudez C, Martinez P, Cadevall C and Artigas P 2019 Active illumination focus variation *Proc. SPIE* **1105** 110560W

Bitte F, Mischo H, Pfeifer T and Frankowski G 2000 3D surface inspection with a DMD based sensor *Proc. IMEKO World Congress 16 (Vienna)* 157–62

Blateyron F 2011 Chromatic confocal microscopy *Optical Measurement of Surface Topography* ed R K Leach (Berlin: Springer)

Dan D, Yao B and Lei M 2014 Structured illumination microscopy for super-resolution and optical sectioning *Chin. Sci. Bull.* **59** 1291–307

Fay M F, Colonna de Lega X and de Groot P 2014 Measuring high-slope and super-smooth optics with high-dynamic range coherence scanning interferometry *OSA Optical Fabrication and Testing* OW1B-3

Giusca C and Leach R K 2012 Calibration of the scales of areal surface topography-measuring instruments: part 1. Measurement noise and residual flatness *Meas. Sci. Technol.* **23** 035008

Giusca C L and Leach R K 2013 Calibration of imaging confocal microscopes for areal surface texture measurement *Measurement Good Practice Guide No. 129* (Teddington: National Physical Laboratory)

Heintzmann R 2006 Structured illumination methods *Handbook of Biological Confocal Microscopy* ed J Pawley (Berlin: Springer)

Helmli F S and Scherer S 2001 Adaptive shape from focus with an error estimation in light microscopy *Proc. 2nd Int. Symp. Image and Signal Processing and Analysis* 188–93

ISO 25178 part 607 2019 *Geometrical Product Specifications (GPS)—Surface Texture: Areal—Part 607; Nominal characteristics of non-contact (confocal microscopy) instruments* (Geneva: International Organization for Standardization)

ISO/DIS 25178 part 700 2020 *Geometrical Product Specifications (GPS)—Surface Texture: Areal—Part 700: Calibration, adjustment and verification of areal topography measuring instruments* (Geneva: International Organization for Standardization)

James B and Pawley 2010 *Handbook of Biological Confocal Microscopy* (Berlin: Springer)

Kieran G L 2001 Natural demodulation of two-dimensional fringe patterns. I. General background of the spiral phase quadrature transform *J. Opt. Soc. Am.* A **18** 1862–70

Krzysztof P, Maciej T and Tomasz T 2014 Optically-sectioned two-show structured illumination microscopy with Hilbert–Huang processing *Opt. Express* **22** 9517–27

Leach R K 2015a Abbe error/offset *CIRP Encyclopaedia of Production Engineering* ed L Laperrière and G Reinhart (Berlin: Springer)

Leach R K 2015b Is one step height enough? *Proc. ASPE (Austin, USA)*

Leach R K 2020 *Advances in Optical Form and Coordinate Metrology* (Bristol: IOP Publishing)

Leach R K, Giusca C L, Haitjema H, Evans C and Jiang X 2015 Calibration and verification of areal surface texture measuring instruments *Ann. CIRP* **64** 797–813

Leach R K, Giusca C and Rubert P 2013 A single set of material measures for the calibration of areal surface topography measuring instruments: the NPL areal bento box *Proc. Met Props. (Taipei)* 403–6

Liu J, Liu C, Tan J, Yang B and Wilson T 2016 Super-aperture metrology: overcoming a fundamental limit in imaging smooth highly curved surfaces *J. Micros.* **261** 300–6

Liu J and Tan J 2016 *Confocal Microscopy* (San Rafael, CA: Morgan and Claypool)

Martinez P, Bermudez C, Cadevall C, Matilla Ayala A and Artigas R 2020 Three-dimensional imaging confocal profiler without in-plane scanning *Proc. SPIE* **11352** 113502L

Martinez P 2018 Residual flatness error correction in three-dimensional imaging confocal microscopes *Master's Thesis* Escuela Técnica Superior de Ingeniería de Telecomunicación de Barcelona https://upcommons.upc.edu/bitstream/handle/2117/123905/TFM_PolMartinez.pdf?isAllowed=y&;sequence=1

Matilla A, Bermudez C, Mariné J, Martínez D, Cadevall C and Artigas R 2017 Confocal unrolled areal measurements of cylindrical surfaces *Proc. SPIE* **10329** 1032915

Matilla A, Mariné J, Pérez L, Cadevall C and Artigas R 2016 Three-dimensional measurements with a novel technique combination of confocal and focus variation with a simultaneous scan *Proc. SPIE* **9890** 98900B

Minsky M 1957 *US Patent* #3013467

Nimishankavi L P, Jones C, O'Connor D and Giusca C L 2019 NPL areal standard: a multi-function calibration artefact for surface topography measuring instruments *Proc. Laser Metrology and Machine Performance XIII* 69–72

Noguchi M and Nayar S K 1994 Microscopic shape from focus using active illumination *Proc. 12th Int. Conf. Pattern Recognit.* 147–52

Ramasamy S K and Raja J 2013 Performance evaluation of multi-scale data fusion methods for surface metrology domain *J. Manuf. Syst.* **32** 514–22

Sheppard C J R and Mao X Q 1988 Confocal microscopes with slit apertures *J. Mod. Opt.* **35** 1169–85

Schwertner M 2011 Method and assembly for optical reproduction with depth discrimination *US Patent* 7.977.625

Wang J, Leach R K and Jian X 2015 Review of the mathematical foundations of data fusion techniques in surface metrology *Surf. Topogr.* **3** 123001

Wicker K and Heintzmann R 2010 Single-shot optical sectioning using polarization-coded structured illumination *J. Opt.* **12** 084010

Xing Z 2015 Double-exposure optical sectioning structured illumination microscopy based on Hilbert transform reconstruction *PLoS One* **1371** 0120892

Zachary R, Hoffman, Kivanc K and DiMarzio C A 2018 Single image structured illumination (SISIM) for *in-vivo* imaging *Proc. SPIE* **10488** 1049918

Zint M, Stock K, Claus D, Graser R and Hibst R 2019 Development and verification of a snap shot dental intra-oral 3D scanner based on chromatic confocal imaging *J. Med. Imaging* **6** 033502

Chapter 5

Non-scanning techniques

Xiaobing Feng, Zhengchun Du and Jianguo Yang

Surface metrology is increasingly employed to perform in-process measurement tasks for the purpose of process monitoring. While techniques such as coherence scanning interferometry and imaging confocal microscopy produce highly accurate topography maps, the measurement speed is often insufficient for in-process measurement due to the mechanical scanning required. Non-scanning techniques, on the other hand, have the potential for high-speed measurement, making them suitable for in-process measurement. This chapter will review four commonly used non-scanning surface texture measurement techniques (wavelength scanning interferometry, dispersed reference interferometry, chromatic confocal microscopy and micro-scale fringe projection). The working principle of each technique will be briefly presented and recent research advances summarised.

5.1 Introduction to non-scanning techniques

5.1.1 Scanning versus non-scanning techniques

The techniques described in previous chapters (chapters 2–4) typically rely on mechanical movement of the sensor head or sample in the axial or surface height direction in order to reconstruct the surface topography. The axial scanning motion is often driven by piezoelectric devices or precision motors, with typical scanning distances from tens of micrometres to a few millimetres. The moving components often include the objective lens, critical components in the optical path (for example, the reference mirror in the case of coherence scanning interferometry and the pinhole in the case of chromatic confocal microscopy) and the image sensor. Alternatively, the optics can remain stationary and the surface being measured is moved. Images are recorded during the axial scanning motion at regular intervals (typical stepping distances range from hundreds of nanometres to a few micrometres) to achieve sufficient axial discrimination in height measurement. The techniques described in this chapter do not require axial scanning in order to measure the surface topography. Since the axial scanning motion is often the most time-consuming part of

doi:10.1088/978-0-7503-2528-8ch5

the measurement process, non-scanning techniques have the potential to be implemented for high-speed surface measurement, which is highly desirable in manufacturing applications where in-process measurement is required (see chapter 7). Another advantage of not requiring axial scanning is that measurement errors associated with mechanical dynamics, such as the linearity of the motion driver and vibration, can be minimised or even avoided.

A potential weakness of non-scanning techniques is rarely reported in the published work and is one of the primary reasons that many commercial systems use physical scanning. This weakness is down to the fact that the mechanisms used to obtain the required data mean that the instruments are essentially working out of focus (Chakmakjian *et al* 1996). This means that the transfer function of the instruments is not as well defined as for a focused system, resulting in difficult-to-predict spatial frequency distortion. Scanning systems, such as coherence scanning interferometry, imaging confocal microscopy and focus variation provide the equivalent of an infinite depth of focus: every topography measurement is taken at the position of best focus. This can significantly improve performance while extending the measurement range of the instrument, particularly at higher magnifications. This effect should be considered when using all of the non-scanning instruments described in this chapter.

5.1.2 Definition of a non-scanning technique

The word 'scanning' is often used to describe different actions during surface measurement. For example, in wavelength scanning interferometry, the scanned variable is the wavelength of the light provided by the source; in coherence scanning interferometry, the scanned variable is the axial distance between the sample and the optical sensor; in lateral scanning it is the sample that is moved in the lateral plane in order to perform profile or areal measurement using a point-measuring or line-measuring instrument. Therefore, it is important to clarify that in this chapter, 'non-scanning' techniques refer to *techniques where there is no relative movement along the optical axis between the sensor head and the sample surface.*

The following sections describe some of the commonly used non-scanning techniques, including wavelength scanning interferometry, dispersed reference interferometry, chromatic confocal microscopy and micro-scale fringe projection. To the best of the author's knowledge, there is currently no published work that provides an in-depth review of these non-scanning techniques together. Therefore, a brief description of the working principles of each technique will be provided along with recent technical advances. The following acronyms will be used throughout the chapter.

AOTF	acousto-optic tunable filter
BS	beam splitter
CCD	charge-coupled device
CCM	chromatic confocal microscopy
CMOS	complementary metal-oxide-semiconductor
CSI	coherence scanning interferometry

DLP digital light processing
DMD digital mirror device
DRI dispersed reference interferometry
FFT fast Fourier transform
FP fringe projection
ICM imaging confocal microscopy
LCoS liquid crystal on silicon
OPD optical path difference
PSI phase shifting interferometry
WSI wavelength scanning interferometry

5.2 Wavelength scanning interferometry

WSI is an interferometric technique where the necessary phase shift in the interfered beams is provided by scanning the wavelength of the light source. WSI is able to measure absolute distances without the 2π phase ambiguity limitation of PSI (de Groot 2011), which enables a larger height measuring range. Additionally, WSI is able to perform faster measurement than CSI, as mechanical axial scanning is not required, thus minimising issues of vibration and nonlinearity in the scanning device.

Figure 5.1 illustrates a typical WSI set-up in a Michelson configuration. The bulk optics interferometer set-up is similar to that of CSI, where the light beam is separated at a beam splitter (BS) and propagates along the measurement and

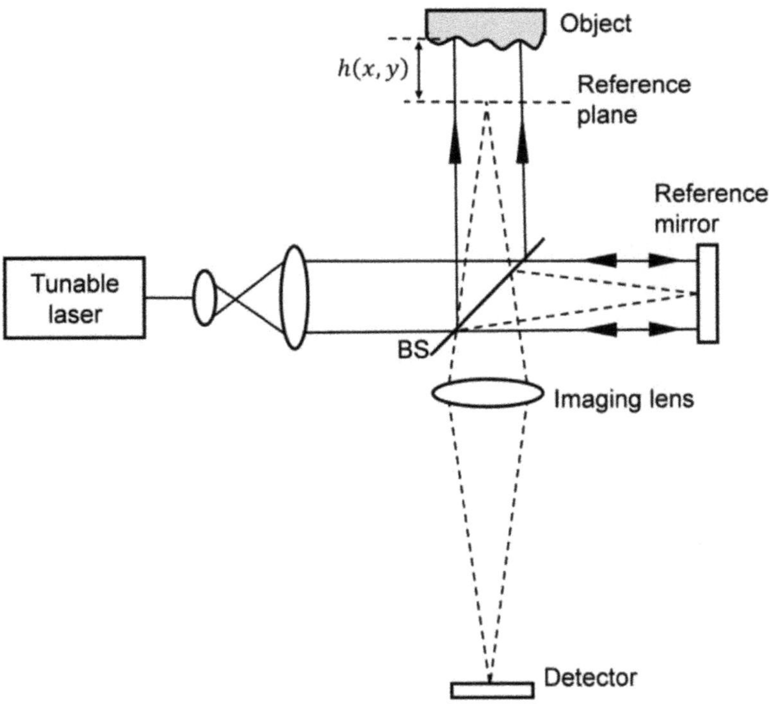

Figure 5.1. Schema of a wavelength scanning interferometer configuration (adapted from Yamamoto *et al* 2001). $h(x,y)$ is the distance between the sample surface and the reference plane.

reference arms. The two beams are recombined at the BS and the interfered beams are received by a detector, such as a CCD camera.

In WSI, a light source with tunable wavelength is employed to emit a sequence of beams with different wavelengths. Each beam in the sequence forms an interference signal which is captured by the detector. Changes in the phase difference between the measurement and reference arms are affected by scanning the light source wavelength, whereas in CSI they are produced by mechanical scanning.

The interference signal of a Michelson interferometer can be expressed as

$$I(x, y) = I_r(x, y) + I_o(x, y) + 2\sqrt{I_r(x, y)I_o(x, y)} \cos [\varphi(x, y)], \tag{5.1}$$

where x and y are the lateral position of the measuring point (corresponding to one pixel in the imaging sensor), I_r and I_o are the intensities of the reference beam and the object beam, respectively, and φ is the phase difference between the two beams, which can be expressed as

$$\varphi = \frac{4\pi}{\lambda}h + \varphi_0, \tag{5.2}$$

where λ is the wavelength of the source light. h is the distance from the reference plane (i.e. the plane where the optical path difference is zero), which corresponds to a half of the OPD, and φ_0 is a phase offset related to the reflection and transmission properties of the interferometer components.

The shift in phase difference $\Delta\varphi$ due to a wavelength shift of $\Delta\lambda$ is expressed as

$$\Delta\varphi = 4\pi h\left(\frac{1}{\lambda} - \frac{1}{\lambda + \Delta\lambda}\right). \tag{5.3}$$

If λ and $\Delta\lambda$ are known, h can be obtained by measuring the shift in the phase differences.

The axial resolution of WSI is determined by the wavelength tuning range and given by

$$\Delta h = \frac{1}{2}\left(\frac{1}{\lambda_{\min}} - \frac{1}{\lambda_{\max}}\right)^{-1}, \tag{5.4}$$

where $\lambda_{\min}$ and $\lambda_{\max}$ are the minimum and maximum wavelengths within the tuning range.

The subsequent phase determination, which leads to a surface height measurement, can be performed with similar methods to those used in CSI, such as using the FFT.

The measurable range in the axial direction is expressed as

$$\Delta H = \frac{\lambda_{\max}\lambda_{\min}}{4\Delta\lambda}. \tag{5.5}$$

Therefore, the measurable range can be increased by decreasing the wavelength tuning step. The measurable range is also limited by other factors, such as the coherence length of the laser and the depth of focus of the imaging system.

5.2.1 Advances in wavelength scanning interferometry

WSI was initially developed for high-precision distance measurement. In 1987, frequency modulated radar technology was applied through adjustment of the laser diode frequency to develop an interferometer that measured relative phase and absolute distance (Kubota *et al* 1987). Following this, an implementation of PSI with a tunable laser was proposed in 1990 to measure surfaces (Okada *et al* 1990).

Earlier implementations of WSI utilised tunable diode laser sources (Kikuta *et al* 1986, Thiel *et al* 1995, Kubota *et al* 1987, Okada *et al* 1990, Ishii 1999), where the wavelength of the emitted light was tuned by modulating the diode's injection current, which can be carried out at gigahertz rates. Axial resolution increases with the wavelength scanning range, indicated by equation (5.4). Tens of micrometres axial resolution can be achieved with tunable laser diodes in which the wavelength tuning range is a few nanometres. Using wavelength tunable dye lasers with a narrow 4.2 nm (Yamaguchi *et al* 2000) and a wider 25 nm (Kuwamura and Yamaguchi 1997) scanning width, 39 and 1 μm axial resolutions were achieved, respectively. Sub-micrometre accuracy was achieved with a semiconductor laser diode with an external cavity (Thiel *et al* 1995), where the tuning range reached tens of nanometres. Further improvement in axial resolution was achieved with wide-range sources, such as solid-state Ti:sapphire lasers (Davila *et al* 2012a) or halogen light sources coupled with an AOTF (Jiang *et al* 2010). A commercial Ti:sapphire laser originally designed for spectroscopic applications was modified to allow a wide scanning range of approximately 100 nm and a picometre level wavelength step for use in WSI, effectively improving axial resolution by a factor of more than 100 (Davila *et al* 2012b).

Yamamoto *et al* (2001) implemented WSI for surface form measurement using an electronically tuned Ti:sapphire laser and high-speed CCD camera. The set-up achieved a field of view of 5.12 mm $\times$ 5.12 mm, a lateral resolution of 10 μm and a measurement range of 1.56 mm. The height measurement error evaluated on a 10 μm step height was 0.1 μm. The measurement resolution was limited by coherent noise due to reflection from the backside of optical components. Yamamoto and Yamaguchi (2002) examined the capability of an implementation with a dye laser for measuring plane surfaces at various slope angles.

A dual interferometer WSI set-up, as shown in figure 5.2, has been developed (Jiang *et al* 2010) to improve surface measurement accuracy by compensating for environmental noise, such as vibration and temperature drift. The interferometer used for surface measurement is illuminated with a halogen lamp. Wavelength scanning is performed with an AOTF, allowing a measurement range of approximately 200 μm. An additional reference interferometer, illuminated by a near-infrared superluminescent light-emitting diode, is used to monitor and compensate for the environmental noise. Light beams from the two sources are coupled to an

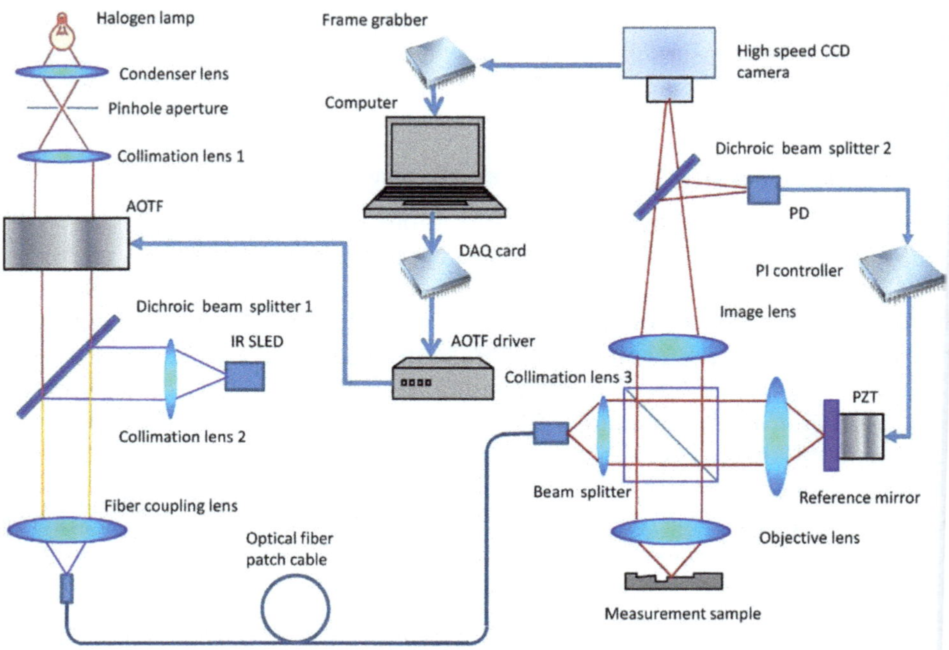

Figure 5.2. Schema of a dual interferometer set-up for environmental noise compensation (reprinted with permission from Jiang *et al* 2010, copyright The Optical Society).

optical fibre and travel almost the same optical path. As a result of the shared optical path, the reference interferometer is able to monitor any optical path change due to the environmental noise and compensate by moving the reference mirror with a piezo-electric translator to cancel out the optical path changes. The instrument was able to achieve 1 nm height measuring error on two step height material measures (2.970 μm and 292 nm). Such a set-up is especially suitable for use in the manufacturing environment, which was demonstrated (Jiang 2011) with an *in situ* measurement of micro-structured surfaces on a large drum diamond turning machine.

The influence of fringe analysis algorithms on the reconstruction accuracy of WSI and computation speed has been examined (Muhamedsalih *et al* 2012). Four algorithms were compared: (i) phase shift determination using simple FFT; (ii) phase shift determination using fitted FFT; (iii) interference pattern analysis using FFT; and (iv) localised peaks of interference pattern using convolution. The latter two algorithms were found to provide more than ten times improvement in height measuring error over the FFT based algorithms. Optimisation of the filter design in existing fringe analysis algorithms has been proposed (Zhang *et al* 2018b), and it was found that with parameter optimisation measurement accuracy can be improved, in particular when the signal-to-noise ratio is low.

A method to double the measurement range of WSI without the need for narrower wavelength scanning steps was proposed (Moschetti *et al* 2016b) by solving the sign ambiguity in the OPD using quadrature WSI. Using *a priori* knowledge of the signal background, only one additional WSI signal is required. In a measurement comparison with standard WSI using a 30 μm step height material measure, the OPD sign ambiguity resulted in incorrect height measurement (measured height = 0.72 μm) with standard WSI, whereas quadrature WSI was not affected (measured height = 29.8 μm). In addition, the quadrature WSI method significantly reduced non-linearities in the instrument's vertical axis response due to the processing algorithm. The mechanical scanning re-introduced in quadrature WSI did not affect the measurement speed, as it was limited by the CCD camera frame rate. The authors (Moschetti *et al* 2016a) also successfully improved the repeatability of WSI, to a level comparable to PSI and CSI, by resolving fringe order ambiguity using fringe frequency and phase information.

The use of WSI for surface measurement includes a range of applications. WSI implemented with a solid-state tunable laser has been demonstrated to be capable of measuring depth-resolved displacement fields through semi-transparent scattering surfaces (Ruiz *et al* 2005). The surface topography and thickness of transparent films (more than 3 μm thickness) have been measured using WSI by applying an optimised Fourier transform method in fringe analysis (Gao *et al* 2012). A dual probe WSI set-up was developed (Zhang *et al* 2018a) to measure in a single acquisition acute-angled vee-groove structures that are challenging to measure due to the high surface slope angles and multiple reflection at the groove bottom. Two WSI probes with orthogonal measurement planes and sharing the same light source were deployed and pre-calibrated for surface topography registration. Comparison with a stylus instrument demonstrated profile measurement deviations within 0.8 μm. Recently, WSI implemented with a Fizeau interferometer has been developed for measuring the surface profiles of transparent plates and achieved profile measurement errors below 2.3 nm on both surfaces (Sun *et al* 2019).

5.3 Dispersed reference interferometry

DRI is a type of interferometry where a wavelength-dependent optical path is generated by introducing chromatic dispersion to the reference arm. Where spectral interferometry is used more widely to describe the spectral analysis of interferograms, DRI specifically refers to the application of chromatic dispersion to the reference field. DRI has also been referred to as dispersive white-light interferometry, white-light interferometry with dispersion, dispersive reference interferometry and dispersion-encoded low-coherence interferometry.

A schema of a typical DRI implementation with a Michelson interferometer is shown in figure 5.3. A broadband light source is collimated before being separated into two arms by the BS. Light in the measurement arm is reflected from the sample surface, while light in the reference arm is reflected by the reference mirror. A dispersive element is placed in the reference arm to introduce a wavelength-dependent optical path length. The two arms are recombined at the BS and received by a spectrometer.

The spectral intensity at the output of the interferometer is given by

$$I(k) = \frac{I_{k0}}{2}[1 + \cos\phi(k)], \tag{5.6}$$

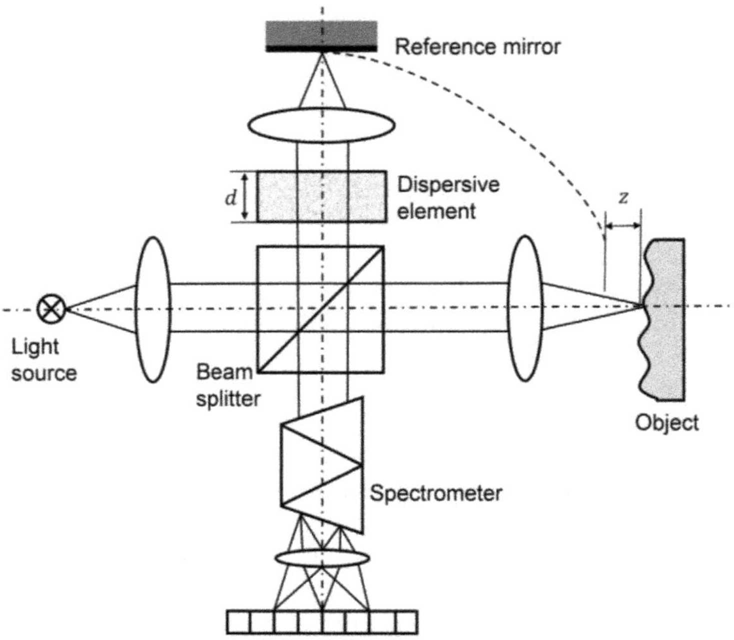

Figure 5.3. Schema of a typical dispersed reference interferometer with a Michelson configuration (adapted from Pavlíček and Häusler 2005).

where k is the wavenumber, k_0 is the central wavenumber of the light source, I_{k0} is the spectral intensity of the light source and ϕ is the phase difference between the two interferometer arms. An example of this spectral intensity is shown in figure 5.4 (top graph).

The relationship between ϕ and the distance between the object surface and the reference plane z can be represented by

$$\phi(k) = 2k[z + d - dn(k)], \tag{5.7}$$

where d and n are the thickness and refractive index of the dispersive element, respectively. An example of this phase difference is shown in figure 5.4 (bottom graph), which is in the form of a parabola, at the vertex of which the interferometer is 'balanced'. The resulting interferogram is symmetrical about the wavenumber associated with the phase vertex (Martin and Jiang 2013). The stationary phase point manifests at an equalisation wavelength at which the interferometer OPD is zero. Surface height can, therefore, be determined using the equalisation wavelength. The height measuring resolution mainly depends on the resolving power of the spectrometer and the algorithm used to determine the equalisation wavelength λ_{eq}. Since DRI requires a spectrometer as the detector, where one dimension of the photodetector array is used to obtain spectral information, it is inherently a point sensing (with a linear detector) or line sensing (with an areal detector) technique. Hence, lateral scanning is required for areal surface measurement.

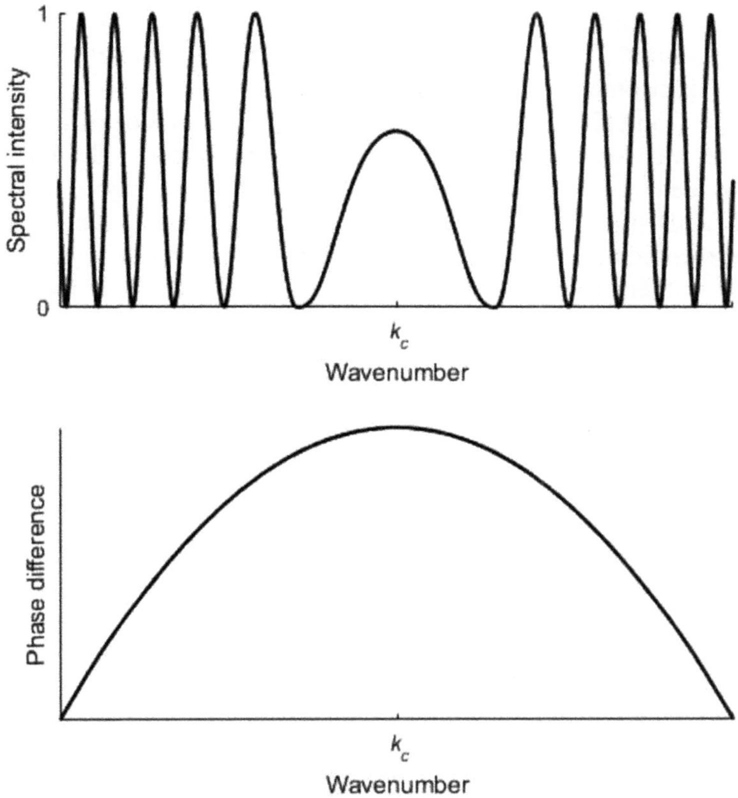

Figure 5.4. Spectral interferogram (top) and phase difference (bottom) against angular wavenumber in DRI. k_c is the wavenumber at which the interferometer is balanced.

5.3.1 Advances in dispersive reference interferometry

Implementations of DRI mainly differ in the dispersive element in the reference arm and the light source. Hlubina (2002) used a quartz tungsten halogen lamp to produce broadband white-light illumination, and an optical sample of known thickness and wavelength-dependent refractive index to affect the OPD. The instrument was used for displacement measurement and was able to achieve better than 1 μm accuracy for a 10 μm displacement. The range of measurable distances depended on the thickness of the optical element.

Pavlíček and Häusler (2005) used a superluminescent diode with central wavelength of 815 nm and a spectral width of 20 nm for illumination, and relied on material dispersion within optical fibres to produce the OPD. An optical fibre with higher dispersion in the reference arm than in the measurement arm is used to utilise the dispersion effect. The instrument was able to achieve a depth measuring range of 0.9 mm. The height measuring repeatability was 0.05 μm in the centre of the measurement range and 0.11 μm at its margin. Surface profiles on two roughness material measures were measured using the set-up, and the roughness parameter

Ra values were computed based on the measured profiles and compared with reference *Ra* values. The measurement speed was mainly limited by the sampling speed of the spectrometer which, in this case, was 80 Hz.

Martin and Jiang (2013) also used a superluminescent diode as the light source, with a central wavelength of 829.2 nm and a spectral width of 19.5 nm. In this set-up, a pair of identical transmission diffraction gratings were placed in the reference arm to affect the OPD. The reference beam was dispersed after traversing one grating and recollimated by the other grating. Only the first diffracted orders were used, whereas other orders were blocked. The spectral interferogram was provided by a grating-based spectrometer detector with 2048 pixels and 12-bit analogue-to-digital conversion resolution. Due to the way dispersion was generated in this set-up, the loss of light intensity in the first diffracted orders led to longer sampling periods in the spectrometer, resulting in a measurement sampling rate of 20 Hz. The instrument achieved a height measuring resolution of 279 nm. The instrument had a repeatability of 197 nm in the worst case and a linearity of 0.51% (i.e. 510 nm) over a 100 μm measurement range. A version of this set-up was subsequently mounted on an ultraprecision turning machine to demonstrate its capability for on-machine surface measurement (Li *et al* 2018). After corrective machining based on on-machine DRI measurement, the profile accuracy of a cosine curve sample was improved from 104.7 to 58.6 nm.

The axial measuring resolution of DRI is low compared to scanning interferometry where distance is encoded in phase information. To overcome this limitation, Williamson *et al* (2016) successfully improved the axial measuring resolution of the instrument implemented by Martin and Jiang (2013) by further retrieving the phase information from the spectral interferogram with template matching. A numerical model of the spectral interferogram was developed to generate a set of template interferograms, which were then matched with the detected interferogram. Displacements of 10 nm could be resolved as a result. The metrological characteristics were also significantly improved over the previous set-up: the measurement noise was 0.63 nm when 1401 interferograms were acquired over a period of 60 s; the repeatability was 1.25 nm and the linearity was 40 nm. The main disadvantage of this method of retrieving phase information is that it requires a lengthy calibration procedure. Henning *et al* (2019) examined the use of a linear approximation in the analysis of spectral interferograms in the previous work of Pavlíček and Häusler (2005) and Martin and Jiang (2013) and found that it could lead to a significant measurement error of 2.78 μm over a measurement range of 290 μm. To correct such errors, the authors presented the full expression to replace the linear approximation when a DRI system is used for high-precision measurement at a sub-micrometre scale.

In more recent research to improve axial measurement precision, Taudt *et al* (2020) proposed a high-dynamic-range design utilising the signal behaviour in which the amplitude of the signal at the equalisation wavelength is sensitive to small height changes. By fitting the signal amplitude in a narrow region of interest near λ_{eq},

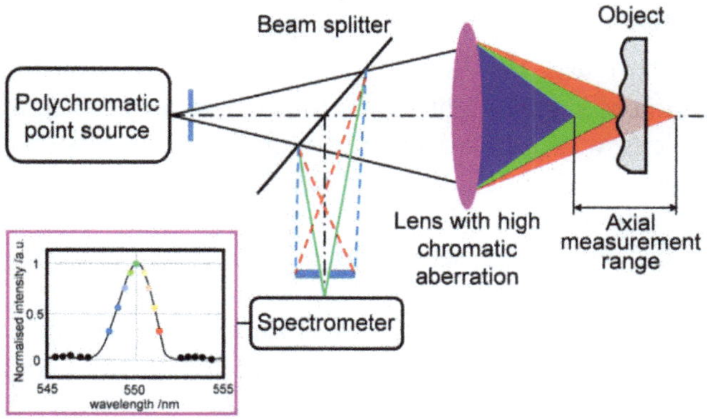

Figure 5.5. Schema of a typical CCM. (Adapted from Chen *et al* 2019b).

repeatability as low as 0.13 nm was achieved on a low scattering 100 nm height material measure. Consequently, axial resolution could be decoupled from axial measurement range, resulting in an axial measurement range/resolution ratio (i.e. dynamic range) of nearly 8×10^5 in the best case. In addition, an imaging lens was actuated to achieve lateral scanning and hence areal surface measurement, without movement of either the sample or the reference mirror. Since the lens was placed after both interferometric arms were combined, measurement was not affected by errors usually associated with the mechanical movement of the sample, which are common when the sample is moved to achieve lateral scanning.

5.4 Chromatic confocal microscopy

CCM is a type of confocal microscopy where chromatic dispersion is used to produce light with various wavelengths for measuring surface height. CCM is able to generate an axial response in a single acquisition and surface height can be determined with spectroscopic analysis of the detected signal. Measurement speed is faster than ICM, where the axial response is affected by vertical scanning (but see chapter 4). Areal surface measurement using CCM is typically performed by lateral scanning or parallel sensing.

Figure 5.5 is a schema of a typical CCM instrument, which includes a co-axial illumination optical set-up similar to that in an ICM, a broadband light source, a spectrometer and a pinhole in front of the spectrometer. A specially designed objective lens or a diffractive lens is used to promote a longitudinal chromatic dispersion effect which results in different spectral components of the broadband light source being focused at different planes along the optical axis. The pinhole is positioned so that light that is best focused onto the sample surface returns to the detector with highest flux, forming a peak at the corresponding wavelength in the detected spectrum. Surface height is, therefore, determined by locating the peak in the spectrum signal.

A more detailed description of the working principles of CCM is available elsewhere (Blateyron 2011).

5.4.1 Advances in chromatic confocal microscopy

Molesini *et al* (1984) was the first to develop an optical profiler that utilised chromatic dispersion of a white-light source to perform surface height measurement. A plano-convex lens was used to create longitudinal dispersion. The returning spectral components were angularly separated with a prism and subsequently detected by a photodiode array. The instrument was able to achieve 0.1 μm height measuring accuracy without the sectioning power of the confocal principle (i.e. a pinhole was not used).

Tiziani and Uhde (1994) demonstrated chromatic sectioning with a confocal microscope using three chromatic filters and a camera with black-and-white film. By mapping the colour tone of the received light to surface height, an axial resolution of 45 nm was realised. The colour tone analysis approach was somewhat different to the spectral analysis approach commonly used in modern CCM.

Earlier CCM implementations were point sensing probes. To obtain areal or profile measurement, transverse scanning or parallel sensing is necessary. Stage scanning is a form of transverse scanning where the beam stays stationary while the object is moved. It is conceptually simple but the mechanical movement significantly limits the measurement speed. Moving pinhole scanning is a faster transverse technique, where the object remains stationary, whereas the position of the light source and detection points are moved, such as with a rotating Nipkow disk (Tiziani and Uhde 1994) or a digital micromirror device (Cha *et al* 2000). However, moving pinhole scanning has the drawback of low light efficiency which leads to higher measurement noise and lower accuracy. With parallel sensing, single shot surface profile measurement is possible using a slit-scan confocal instead of a pinhole set-up (Lin *et al* 1998), with a laterally segmented diffractive lens combined with spectral multiplexing (Hillenbrand *et al* 2013), or with time encoded spectroscopy (Park *et al* 2019). Direct areal surface measurement has also been achieved with diffractive lenses, such as a 100 × 100 microlens array as the objective lens employed by Tiziani *et al* (1996). However, the lateral sampling distance provided by diffractive lenses is limited by the size of the individual elements and their fabrication accuracy. Another type of transverse scanning technique is beam scanning (Chun *et al* 2009), where the incident angle of the beam is changed with devices, such as galvanometer mirrors, rotating polygonal mirrors and acousto-optic deflectors. A schema of a galvanometer mirror-based beam scanning CCM set-up is shown in figure 5.6. Two galvanometer mirrors are placed in the optical path with orthogonal rotating axes, such that one mirror steers the collimated light along the x-axis on the sample, and the other mirror steers the light along the y-axis. Beam scanning can be performed at sufficiently high speed, with diffraction limited lateral resolution and high light efficiency.

The axial measurement range of CCM largely depends on the spectral range of the employed light source. Conventional white-light sources (Lin *et al* 1998) with wide spectral ranges, such as xenon lamps, provide a wide measurement range, but the low spatial coherence of white-light sources and the instability of their spectra, often result in low illumination efficiency and limit the measurement accuracy

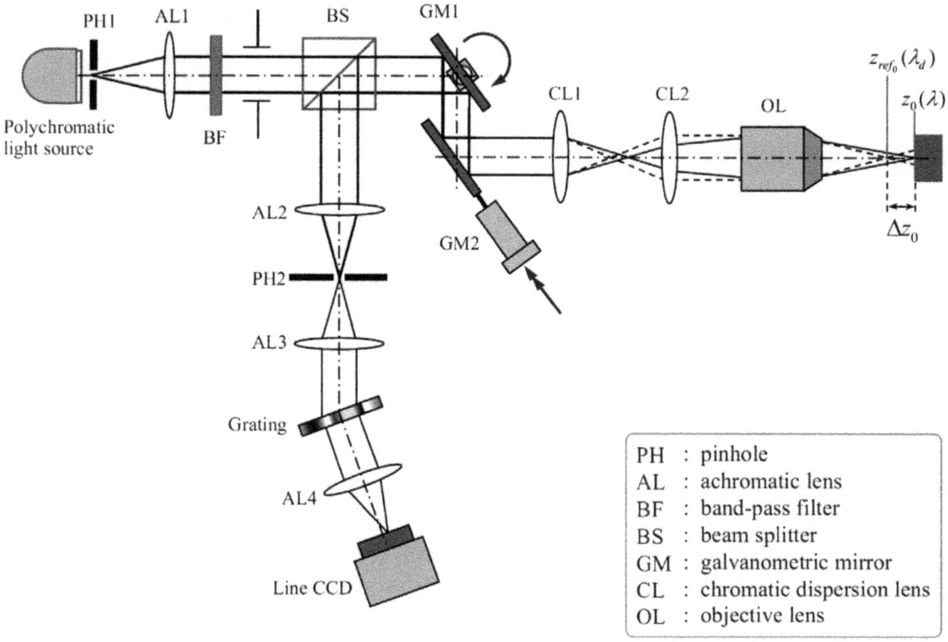

Figure 5.6. Schema of a beam-scanning chromatic confocal microscope (reprinted from Chun *et al* 2009 with the permission of AIP Publishing).

(Chen *et al* 2018). A supercontinuum laser with large bandwidth and high spatial coherence was utilised by Shi *et al* (2004) in CCM, where a measurement range of 7 μm and axial measuring resolution better than 1 μm was achieved. Due to high illumination efficiency, the signal-to-noise ratio was 1000 times better than typically achievable with xenon lamps, enabling significantly higher measurement speed. A mode-locked femtosecond laser with a highly stable optical spectrum and high spatial coherence was employed by Chen *et al* (2018) in a differential confocal set-up. The spectral non-smoothness of the laser source was eliminated with a dual-detector to acquire two confocal signals, one at focus and the other out of focus, which effectively expands the working spectral range from 50 nm to the full spectrum width of 180 nm. A height measuring range of 39.3 μm was achieved with an axial measuring resolution better than 30 nm.

Measurement accuracy in CCM depends highly on the methods employed to extract surface height from the detected spectrum. Several algorithms have been utilised for peak extraction (Ruprecht *et al* 2002, Chen *et al* 2019b, Bai *et al* 2019, Lu *et al* 2019, Chen *et al* 2019a, Luo *et al* 2012, Tan *et al* 2015) capable of achieving subpixel wavelength resolution of the spectrometer. Since CCM and ICM produce similar axial response signals, the peak extracting algorithms used are often similar. Ruprecht *et al* (2002) demonstrated that the centre of gravity algorithm is robust and fast, however, systematic error was present if only a small number of slices is available. Luo *et al* (2012) proposed a Gaussian fitting model to locate the peak wavelength of the recorded spectrum and achieved 0.1 μm axial resolution. Tan *et al* (2015) developed a sinc^2 fitting algorithm based on the ideal shape of the axial response curve, which was shown to yield lower measurement uncertainty than Gaussian fitting. Chen *et al* (2019a) proposed a corrected parabolic fitting method which showed significant improvement in accuracy over conventional parabolic fitting, and moderate improvement over Gaussian and sinc^2 fitting. Bai *et al* (2019) developed a modified centroid algorithm with several virtual pixels interpolated into the pixel-to-intensity spectrum, which improved the focal wavelength measurement fluctuation range. Chen *et al* (2019b) further developed a corrected fitting differential algorithm, which was modified from an efficient linear fitting of the differential signal with an effective error compensation, to achieve 100 times computational efficiency improvement with height extraction accuracy comparable to Gaussian fitting.

In addition to applications in measuring biological tissue (Olsovsky *et al* 2013, Garzón *et al* 2008) and transparent objects (Yu *et al* 2018, Miks *et al* 2010), CCM has been applied to measure surface topography in various applications. A commercially available CCM probe with five-axis motion has been developed for on-machine measurement of freeform and conformal optics (DeFisher *et al* 2011). Similarly, CCM probes have been integrated into machine tools and used for surface quality control (Minoni and Cavalli 2008) and grinding process monitoring (Keferstein *et al* 2008). A beam-scanning CCM probe combined with a position-sensitive detector has been developed to measure nanoscale surface topography (Zhuo *et al* 2018). The metrological performance of a commercial CCM surface topography measuring instrument have been evaluated (Seppä *et al* 2018).

Compared to the two spectrum-based interferometric techniques described in sections 5.2 and 5.3, CCM does not obtain phase information useful for high resolution distance measurement; the height measuring resolution achievable with CCM is generally lower than that with WSI and DRI.

5.5 Micro-scale fringe projection

FP is a method of measuring the object surface by projecting light patterns, typically a series of straight lines, onto the surface and analysing the distortion of the patterns due to the surface shape. Given that the principle of measurement is triangulation, axial resolution is expected to be inferior to WSI, DRI and CCM. FP is most commonly used to measure shape and form. However, with high resolution projectors and cameras, FP can be applied at the micro-scale for surface profile measurement. This chapter focuses mainly on research advances relevant to micro-scale FP applied to surface texture measurement. Therefore, some of the reviewed methods may have been developed originally for form measurement but were adopted and applied to surface texture measurement.

Figure 5.7 is a schema of a typical FP instrument, including a projector and a camera viewing from a perspective other than that of the projector. The projector illuminates the object surface with predetermined fringe patterns typically generated through either digital spatial light modulation based on technologies such as DLP, transmissive liquid crystal and LCoS, or laser interference. The camera captures the distorted light patterns, which are analysed to obtain depth information for each

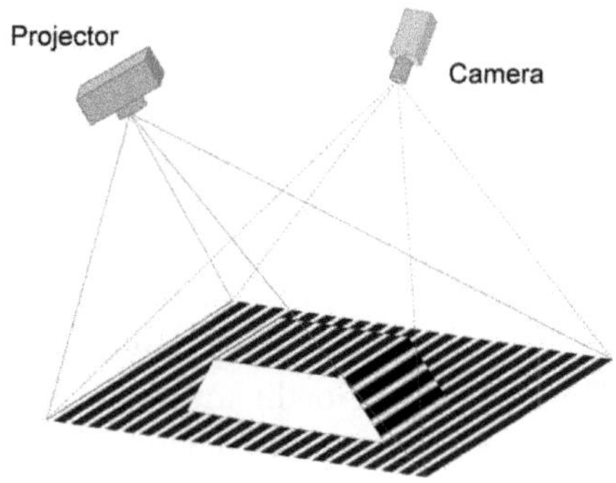

Figure 5.7. Schema of an FP instrument. (Adapted from Van der Jeught and Dirckx 2016.)

pixel in the image, hence producing the areal surface topography. Micro-scale FP often utilises a microscope or high magnification zoom lens to achieve the spatial resolution required for surface measurement.

Various methods have been developed for the generation and analysis of the fringe patterns. An example is the phase shifting technique, a frequently used fringe analysis technique that is well known for high depth accuracy. A sinusoidal fringe pattern is projected onto the surface. The captured image intensity can be described as

$$I_1(x, y) = I'(x, y) + I''(x, y) \cos \phi(x, y), \tag{5.8}$$

where I_1 is the intensity distribution, I' is the DC component, I'' is half of the peak-to-valley intensity modulation and ϕ is the unknown phase.

By generating three additional fringe patterns with phase shifts $\pi/2$, π and $3\pi/2$, additional images are captured with intensities described as

$$I_2(x, y) = I'(x, y) + I''(x, y) \cos \left[\phi(x, y) + \frac{\pi}{2}\right], \tag{5.9}$$

$$I_3(x, y) = I'(x, y) + I''(x, y) \cos [\phi(x, y) + \pi], \tag{5.10}$$

$$I_4(x, y) = I'(x, y) + I''(x, y) \cos \left[\phi(x, y) + \frac{3\pi}{2}\right]. \tag{5.11}$$

The phase $\phi(x,y)$ can be obtained by solving the above four equations, thus

$$\phi(x, y) = \arctan \frac{I_4(x, y) - I_2(x, y)}{I_1(x, y) - I_3(x, y)}. \tag{5.12}$$

The obtained phase is known as the 'wrapped phase' as the calculation produces values ranging from $-\pi$ to π, which need to be unwrapped by correcting the discontinuities in phase. The unwrapped phase can then be used to determine surface depth. The height can be determined as

$$h(x, y) = \frac{\lambda\phi(x, y)}{2\pi}, \tag{5.13}$$

where h is the distance between the reference wavefront and the testing wavefront, which is related to the distance between the surface and the reference plane, and λ is the wavelength of the light. Conversion from phase map to height map can be achieved through calibration (Zhang 2018, Hu *et al* 2003).

Other methods of projection and fringe analysis are widely available and summarised elsewhere (Zhang 2018, Gorthi and Rastogi 2010, Geng 2011, Zuo *et al* 2018, Zhang 2010). Recent advances in FP can also be found elsewhere (Chen *et al* 2020).

5.5.1 Advances in micro-scale FP

Micro-scale FP often requires the combination of the 3D measurement capability of an FP instrument with the resolving power of a microscope, and various configurations have been proposed. One configuration involves bringing the microscope to the FP, hence replacing the regular lenses in an FP instrument with microscope objectives, such as the one developed by Quan *et al* (2002) for surface measurement. Long working distance microscope lenses were placed in front of the projector and the camera. Fringe patterns were generated using a grating with density up to 100 lines per millimetre. Although quantified spatial resolution was not given, standard heights as small as 1 μm were resolved with the device. The measured surface topography of a 0.1 mm × 0.1 mm micromirror suggested lateral resolutions of approximately 2 μm and sub-micrometre axial resolution. An alternative micro-scale FP configuration involves bringing the FP to the microscope, such as that proposed by Zhang *et al* (2002), where a projector and a camera were integrated into a stereomicroscope. The optical configuration of such an implementation is illustrated in figure 5.8. BSs were placed in the optical path before each of the eyepieces, introducing fringe pattern illumination through one path, and capturing the distorted patterns through the other. In addition, fringe patterns were generated using a DMD, which provides faster projection rates and more pattern freedom than using a grating. Measurement of a step height with a nominal height of 23.3 μm and reported height tolerance of 0.14 μm produced a mean step height of 22.3 μm after calibration and error compensation. The measurement error was attributed to the limited positioning precision of the vertical stage during error compensation and the adoption of linear interpolation in phase-to-height conversion.

A similar configuration was developed by Proll *et al* (2003) where the eyepieces were replaced by the projector and the camera. Employing an LCoS display for projection and a CMOS camera, the instrument achieved up to 0.085 μm axial resolution with a field of view of 1.04 mm × 0.77 mm. Li *et al* (2013) characterised a prototype micro-scale FP instrument with a similar configuration to that from Proll *et al* (2003), which achieved a lateral resolution of approximately 3 μm and axial resolution of approximately 1 μm.

To reduce noise in the projected fringe patterns, Thakur *et al* (2007) implemented FP with a grating projection system based on the self-imaging (the Lau effect) of the

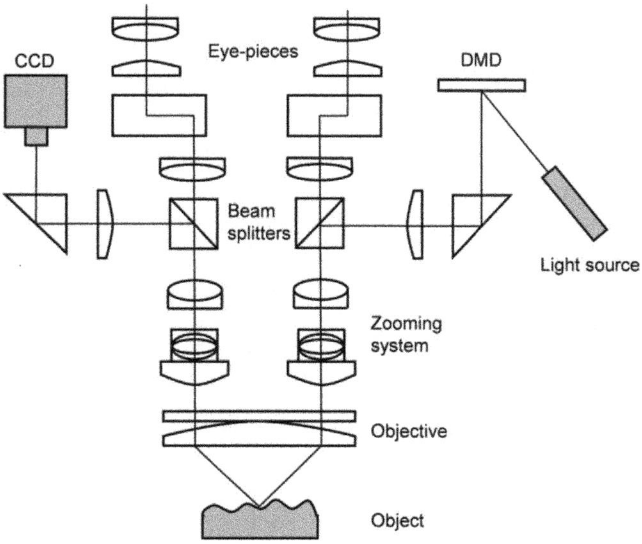

Figure 5.8. Schema of a micro-scale fringe projection instrument with a projector and a camera integrated into a stereomicroscope. (Adapted from Zhang *et al* 2002).

grating. The use of spatially incoherent white-light source and the Lau effect reduced coherent noise and improved the signal-to-noise ratio for surface measurement. Measured surface profiles of a 4 mm height spherical cap and a coin were shown to be within 1.7% and 1.8% average discrepancy with those measured by a stylus instrument.

In order to harvest the high spatial resolution enabled by the phase shifting technique while avoiding its sensitivity to object motion during temporal phase shifting, Chen *et al* (2015) proposed a snapshot phase shifting FP implementation for surface measurement. The illuminating light was split into two orthogonal circularly polarised beams by a combination of polarising BSs, linear polarisers and a quarter wave plate. A four-channel division of focal plane polarisation camera was used to capture four phase shifted images simultaneously. As a result, phase shifts were captured in a single image frame, significantly improving robustness against vibration during surface measurement. Potential error sources were identified and analysed, although no metrological characteristics were provided.

Fringe patterns suitable for the purpose of micro-scale surface measurement have been examined by Li and Zhang (2017), where binary defocusing and sinusoidal projection methods were tested on two microscopic FP implementations. One implementation used telecentric lenses on both the projector and the camera, while the other used a telecentric lens only on the camera. A binary defocusing technique was found to provide up to 19% better axial resolution than sinusoidal FP due to lower noise level, when no image averaging was performed. However, the advantage in resolution narrowed if image averaging was performed. The high projection speed of the binary defocusing technique also enabled microscopic surface measurement at a speed of 500 Hz. Another implementation using binary FP was able to achieve a measurement speed of 120 Hz with a measurement range of 8 mm × 6 mm in lateral and 8 mm in depth (Hu *et al* 2017).

Local specular reflection from surfaces is a common issue affecting surface measurement with FP, given the active illumination required. Differences in the surface texture or local surface slope often result in saturation of the captured images, leading to lower measurement accuracy and non-measured areas. Saturation can be avoided by adaptively adjusting fringe intensity in saturated regions (Liu *et al* 2020), applying multi-polarisation FP (Salahieh *et al* 2014) and employing multi-frequency fringe patterns (Hu *et al* 2019, Tang and Gu 2020). As a result, high-dynamic-range surface measurement can be achieved by micro-scale FP.

In addition to specular reflection, limited depth of focus and occlusions are common issues when measuring the surfaces of objects with complex shape using microscopic lenses. To address these issues, a multi-view micro-scale FP implementation has been developed (Wang *et al* 2017, Yin *et al* 2015) using four cameras, as shown in figure 5.9. The view angles of the cameras were arranged following the Scheimpflug principle to alleviate the limitation on the shared depth of focus and increase vertical measurement range. The implemented instrument was successful in extending the measurement range in depth and reducing occlusions.

FP is a technique developed mainly for object shape measurement. Although micro-scale FP measurement can be performed by employing microscopic lenses, the triangulation principle used for height determination results in limitations in the axial measurement resolution. As a result, micro-scale FP is mainly applied for surface profiling and rarely used for surface texture measurement. While resolution-wise, micro-scale FP is inferior to WSI, DRI and CCM, it has the advantage of full-field surface measurement at high speed, which is promising in the measurement of a small object in motion or dynamic micro-deformation.

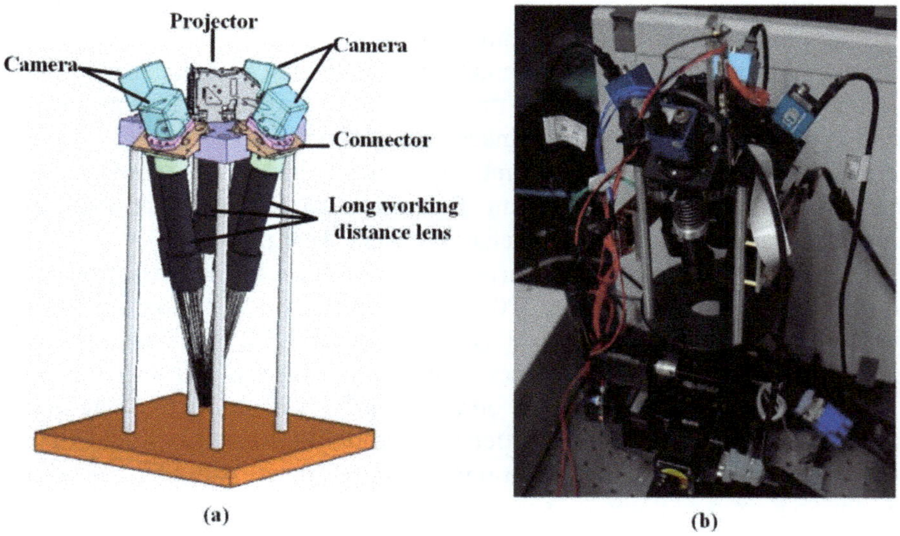

Figure 5.9. A multi-view micro-scale fringe projection instrument (reprinted with permission from Wang *et al* 2017, copyright The Optical Society): (a) illustration and (b) instrument.

References

Bai J, Li X, Wang X, Zhou Q and Ni K 2019 Chromatic confocal displacement sensor with optimized dispersion probe and modified centroid peak extraction algorithm *Sensors* **19** 3592

Blateyron F 2011 Chromatic confocal microscopy *Optical Measurement of Surface Topography* ed R K Leach (Berlin: Springer)

Cha S, Lin P C, Zhu L, Sun P-C and Fainman Y 2000 Nontranslational three-dimensional profilometry by chromatic confocal microscopy with dynamically configurable micromirror scanning *Appl. Opt.* **39** 2605–13

Chakmakjian S, Biegen J F and de Groot P 1996 Simultaneous focus and coherence scanning in interference microscopy *Technical Digest, International Workshop on Interferometry, (Riken, Japan)* 171–2

Chen C, Wang J, Leach R, Lu W, Liu X and Jiang X 2019a Corrected parabolic fitting for height extraction in confocal microscopy *Opt. Express* **27** 3682–97

Chen C, Yang W, Wang J, Lu W, Liu X and Jiang X 2019b Accurate and efficient height extraction in chromatic confocal microscopy using corrected fitting of the differential signal *Precis. Eng.* **56** 447–54

Chen R, Xu J and Zhang S 2020 Digital fringe projection profilometry *Advances in Optical Form and Coordinate Metrology* ed R K Leach (Bristol: IOP Publishing)

Chen X, Nakamura T, Shimizu Y, Chen C, Chen Y-L, Matsukuma H and Gao W 2018 A chromatic confocal probe with a mode-locked femtosecond laser source *Opt. Laser Technol.* **103** 359–66

Chen Z, Wang X and Liang R 2015 Snapshot phase shift fringe projection 3D surface measurement *Opt. Express* **23** 667–73

Chun B S, Kim K and Gweon D 2009 Three-dimensional surface profile measurement using a beam scanning chromatic confocal microscope *Rev. Sci. Instrum.* **80** 073706

Davila A, Huntley J M, Pallikarakis C, Ruiz P D and Coupland J M 2012a Simultaneous wavenumber measurement and coherence detection using temporal phase unwrapping *Appl. Opt.* **51** 558–67

Davila A, Huntley J M, Pallikarakis C, Ruiz P D and Coupland J M 2012b Wavelength scanning interferometry using a Ti:sapphire laser with wide tuning range *Opt. Laser. Eng.* **50** 1089–96

de Groot P 2011 Phase shifting interferometry *Optical Measurement of Surface Topography* ed R Leach (Berlin: Springer)

DeFisher S, Bechtold M and Mohring D 2011 A non-contact surface measurement system for freeform and conformal optics *Proc. SPIE* **8016** 80160W

Gao F, Muhamedsalih H and Jiang X 2012 Surface and thickness measurement of a transparent film using wavelength scanning interferometry *Opt. Express* **20** 21450–6

Garzón J, Gharbi T and Meneses J 2008 Real time determination of the optical thickness and topography of tissues by chromatic confocal microscopy *J. Opt.* A **10** 104028

Geng J 2011 Structured-light 3D surface imaging: a tutorial *Adv. Opt. Photon* **3** 128–60

Gorthi S S and Rastogi P 2010 Fringe projection techniques: whither we are? *Opt. Laser. Eng.* **48** 133–40

Henning A J, Williamson J, Martin H and Jiang X 2019 Improvements to dispersed reference interferometry: beyond the linear approximation *Appl. Opt.* **58** 131–6

Hillenbrand M, Lorenz L, Kleindienst R, Grewe A and Sinzinger S 2013 Spectrally multiplexed chromatic confocal multipoint sensing *Opt. Lett.* **38** 4694–7

Hlubina P 2002 Dispersive white-light spectral interferometry to measure distances and displacements *Opt. Commun.* **212** 65–70

Hu Q, Huang P, Fu Q and Chiang F-P 2003 Calibration of a three-dimensional shape measurement system *Opt. Eng.* **42** 482–6

Hu Y, Chen Q, Feng S, Tao T, Li H and Zuo C 2017 Real-time microscopic 3D shape measurement based on optimized pulse-width-modulation binary fringe projection *Meas. Sci. Technol.* **28** 075010

Hu Y, Chen Q, Liang Y, Feng S, Tao T and Zuo C 2019 Microscopic 3D measurement of shiny surfaces based on a multi-frequency phase-shifting scheme *Opt. Laser. Eng.* **122** 1–7

Ishii Y 1999 Wavelength-tunable laser-diode interferometer *Opt. Rev.* **6** 273–83

Jiang X 2011 *In situ* real-time measurement for micro-structured surfaces *Ann. CIRP* **60** 563–6

Jiang X, Wang K, Gao F and Muhamedsalih H 2010 Fast surface measurement using wavelength scanning interferometry with compensation of environmental noise *Appl. Opt.* **49** 2903–9

Keferstein C P, Honegger D, Thurnherr H and Gschwend B 2008 Process monitoring in non-circular grinding with optical sensor *Ann. CIRP* **57** 533–6

Kikuta H, Iwata K and Nagata R 1986 Distance measurement by the wavelength shift of laser diode light *Appl. Opt.* **25** 2976–80

Kubota T, Nara M and Yoshino T 1987 Interferometer for measuring displacement and distance *Opt. Lett.* **12** 310–2

Kuwamura S and Yamaguchi I 1997 Wavelength scanning profilometry for real-time surface shape measurement *Appl. Opt.* **36** 4473–82

Li A, Peng X, Yin Y, Liu X, Zhao Q, Körner K and Osten W 2013 Fringe projection based quantitative 3D microscopy *Optik* **124** 5052–6

Li B and Zhang S 2017 Microscopic structured light 3D profilometry: binary defocusing technique vs sinusoidal fringe projection *Opt. Laser. Eng.* **96** 117–23

Li D, Jiang X, Tong Z and Blunt L 2018 Development and application of interferometric on-machine surface measurement for ultraprecision turning process *J. Manuf. Sci. Eng.* **141** 014502

Lin P C, Sun P-C, Zhu L and Fainman Y 1998 Single-shot depth-section imaging through chromatic slit-scan confocal microscopy *Appl. Opt.* **37** 6764–70

Liu Y, Fu Y, Cai X, Zhong K and Guan B 2020 A novel high dynamic range 3D measurement method based on adaptive fringe projection technique *Opt. Laser. Eng.* **128** 106004

Lu W, Chen C, Zhu H, Wang J, Leach R, Liu X, Wang J and Jiang X 2019 Fast and accurate mean-shift vector based wavelength extraction for chromatic confocal microscopy *Meas. Sci. Technol.* **30** 115104

Luo D, Kuang C and Liu X 2012 Fiber-based chromatic confocal microscope with Gaussian fitting method *Opt. Laser Technol.* **44** 788–93

Martin H and Jiang X 2013 Dispersed reference interferometry *Ann. CIRP* **62** 551–4

Miks A, Novak J and Novak P 2010 Analysis of method for measuring thickness of plane-parallel plates and lenses using chromatic confocal sensor *Appl. Opt.* **49** 3259–64

Minoni U and Cavalli F 2008 Surface quality control device for on-line applications *Measurement* **41** 774–82

Molesini G, Pedrini G, Poggi P and Quercioli F 1984 Focus-wavelength encoded optical profilometer *Opt. Commun.* **49** 229–33

Moschetti G, Forbes A, Leach R K, Jiang X and O'Connor D 2016a Phase and fringe order determination in wavelength scanning interferometry *Opt. Express* **24** 8997–9012

Moschetti G, Forbes A, Leach R K, Jiang X and O'Connor D 2016b Quadrature wavelength scanning interferometry *Appl. Opt.* **55** 5332–40

Muhamedsalih H, Gao F and Jiang X 2012 Comparison study of algorithms and accuracy in the wavelength scanning interferometry *Appl. Opt.* **51** 8854–62

Okada K, Sakuta H, Ose T and Tsujiuchi J 1990 Separate measurements of surface shapes and refractive index inhomogeneity of an optical element using tunable-source phase shifting interferometry *Appl. Opt.* **29** 3280–5

Olsovsky C, Shelton R, Carrasco-Zevallos O, Applegate B E and Maitland K C 2013 Chromatic confocal microscopy for multi-depth imaging of epithelial tissue *Biomed. Opt. Express* **4** 732–40

Park S J, Jang H and Kim C-S 2019 Time encoded chromatic confocal microscopy for wide field 3D surface profiling *Proc. Eur. Conf. Biomed. Opt.* **11076** 110761P

Pavlíček P and Häusler G 2005 White-light interferometer with dispersion: an accurate fiber-optic sensor for the measurement of distance *Appl. Opt.* **44** 2978–83

Proll K-P, Nivet J-M, Körner K and Tiziani H J 2003 Microscopic three-dimensional topometry with ferroelectric liquid-crystal-on-silicon displays *Appl. Opt.* **42** 1773–8

Quan C, Tay C J, He X Y, Kang X and Shang H M 2002 Microscopic surface contouring by fringe projection method *Opt. Laser Technol.* **34** 547–52

Ruiz P D, Huntley J M and Wildman R D 2005 Depth-resolved whole-field displacement measurement by wavelength-scanning electronic speckle pattern interferometry *Appl. Opt.* **44** 3945–53

Ruprecht A K, Wiesendanger T F and Tiziani H J 2002 Signal evaluation for high-speed confocal measurements *Appl. Opt.* **41** 7410–5

Salahieh B, Chen Z, Rodriguez J J and Liang R 2014 Multi-polarization fringe projection imaging for high dynamic range objects *Opt. Express* **22** 10064–71

Seppä J, Niemelä K and Lassila A 2018 Metrological characterization methods for confocal chromatic line sensors and optical topography sensors *Meas. Sci. Technol.* **29** 054008

Shi K, Li P, Yin S and Liu Z 2004 Chromatic confocal microscopy using supercontinuum light *Opt. Express* **12** 2096–101

Sun T, Zheng W, Yu Y, Asundi A K and Valyukh S 2019 Determination of surface profiles of transparent plates by means of laser interferometry with wavelength tuning *Opt. Laser. Eng.* **115** 59–66

Tan J, Liu C, Liu J and Wang H 2015 Sinc^2 fitting for height extraction in confocal scanning *Meas. Sci. Technol.* **27** 025006

Tang S and Gu F 2020 Adaptive microphase measuring profilometry for three-dimensional shape reconstruction of a shiny surface *Opt. Eng.* **59** 014104

Taudt C, Nelsen B, Baselt T, Koch E and Hartmann P 2020 High-dynamic-range areal profilometry using an imaging, dispersion-encoded low-coherence interferometer *Opt. Express* **28** 17320–33

Thakur M, Quan C and Tay C J 2007 Surface profiling using fringe projection technique based on Lau effect *Opt. Laser Technol.* **39** 453–9

Thiel J, Pfeifer T and Hartmann M 1995 Interferometric measurement of absolute distances of up to 40 m *Measurement* **16** 1–6

Tiziani H J, Achi R and Krämer R N 1996 Chromatic confocal microscopy with microlenses *J. Mod. Opt.* **43** 155–63

Tiziani H J and Uhde H M 1994 Three-dimensional image sensing by chromatic confocal microscopy *Appl. Opt.* **33** 1838–43

Van der Jeught S and Dirckx J J J 2016 Real-time structured light profilometry: a review *Opt. Laser. Eng.* **87** 18–31

Wang M, Yin Y, Deng D, Meng X, Liu X and Peng X 2017 Improved performance of multi-view fringe projection 3D microscopy *Opt. Express* **25** 19408–21

Williamson J, Martin H and Jiang X 2016 High resolution position measurement from dispersed reference interferometry using template matching *Opt. Express* **24** 10103–14

Yamaguchi I, Yamamoto A and Yano M 2000 Surface topography by wavelength scanning interferometry *Opt. Eng.* **39** 40–6 7

Yamamoto A, Kuo C-C, Sunouchi K, Wada S, Yamaguchi I and Tashiro H 2001 Surface shape measurement by wavelength scanning interferometry using an electronically tuned Ti sapphire laser *Opt. Rev.* **8** 59–63

Yamamoto A and Yamaguchi I 2002 Profilometry of sloped plane surfaces by wavelength scanning interferometry *Opt. Rev.* **9** 112–21

Yin Y, Wang M, Gao B Z, Liu X and Peng X 2015 Fringe projection 3D microscopy with the general imaging model *Opt. Express* **23** 6846–57

Yu Q, Zhang K, Cui C, Zhou R, Cheng F, Ye R and Zhang Y 2018 Method of thickness measurement for transparent specimens with chromatic confocal microscopy *Appl. Opt.* **57** 9722–8

Zhang C, Huang P S and Chiang F-P 2002 Microscopic phase-shifting profilometry based on digital micromirror device technology *Appl. Opt.* **41** 5896–904

Zhang S 2010 Recent progresses on real-time 3D shape measurement using digital fringe projection techniques *Opt. Laser. Eng.* **48** 149–58

Zhang S 2018 *High-Speed 3D Imaging with Digital Fringe Projection Techniques* (Boca Raton, FL: CRC Press)

Zhang T, Gao F, Martin H and Jiang X 2018a A method for inspecting near-right-angle V-groove surfaces based on dual-probe wavelength scanning interferometry *Int. J. Adv. Manuf. Technol.* **104** 1–7

Zhang T, Gao F, Muhamedsalih H, Lou S, Martin H and Jiang X 2018b Improvement of the fringe analysis algorithm for wavelength scanning interferometry based on filter parameter optimization *Appl. Opt.* **57** 2227–34

Zhuo G-Y, Hsu C-H, Wang Y-H and Chan M-C 2018 Chromatic confocal microscopy to rapidly reveal nanoscale surface/interface topography by position-sensitive detection *Appl. Phys. Lett.* **113** 083106

Zuo C, Feng S, Huang L, Tao T, Yin W and Chen Q 2018 Phase shifting algorithms for fringe projection profilometry: a review *Opt. Laser. Eng.* **109** 23–59

IOP Publishing

Advances in Optical Surface Texture Metrology

Richard Leach

Chapter 6

Scattering approaches

Mingyu Liu

Light scattering techniques have been widely applied for surface topography measurement, from the measurement of surface texture to the measurement of critical dimensions. With the growing demands from manufacturing industries, advanced light scattering techniques have been developed to address the challenges of fast, in-process, accurate and robust surface topography measurement. Numerous novel applications from industry have also raised different requirements. To meet these requirements, novel systems and algorithms have been developed which have integrated new hardware and software, as well as fast-developing accelerated computational techniques. Applications for mechanical engineering and semiconductor manufacturing, such as scatterometers using extreme ultraviolet lithography technology, defect detection for optical and microstructured surfaces and monitoring of tool wear, have made use of the advantages of light scattering techniques. In addition, emerging technologies in artificial intelligence and hardware acceleration have been used for solving complex inverse scattering problems. This chapter reviews the novel designs of advanced systems and methods using light scattering approaches.

6.1 Introduction

Light scattering is a non-contact, fast and robust method for surface topography inspection. The light scattering technique has attracted a great deal of research attention and there is a long history of using light scattering methods for surface analysis (Beckmann and Spizzichino 1987, Ogilvy 1991, Bennett and Mattsson 1999). The ultimate aim when using a light scattering technique for surface measurement is to fully reconstruct the surface micro-scale topography. However, as the light scattering process is highly non-linear, different surface topographies can have similar light scattering patterns, so fully solving the inverse problem to reconstruct the surface topography can be difficult and is often impossible. The most well-established methods using the light scattering technique are statistical in

doi:10.1088/978-0-7503-2528-8ch6

nature and known as area-averaging methods to determine the surface texture (Leach 2011, Osten 2016). In addition to statistical analysis for surface texture measurement, light scattering has also been applied for critical dimension measurement (CDM), which is widely used in the semiconductor industry (Diebold 2001). Light scattering techniques have also been applied to measure task-specific surface characteristics, such as applications for surface defect detection (Dong *et al* 2019) and tool wear monitoring (Hocheng *et al* 2018). There are applications using light scattering techniques for the detection of food properties (Lu 2016) and detection of the surface quality of pharmaceutical tablets (Bawuah *et al* 2017). This chapter will only focus on light scattering techniques developed in the manufacturing industry.

Systems using light scattering techniques often contain three main components: a light source to project the incident light onto the target surface, a sensor to collect the light scattered from the surface and an analysis unit to process the captured scattered signals to extract useful surface information. According to different applications, each of the three components can have different designs. For example, the light source can be coherent or incoherent, with different wavelengths or different polarisation settings, and fixed or rotatable incident angles; the sensor can be a single-pixel (for example a photodiode) or a sensor array (for example CCD/CMOS), fixed angle or movable, and the captured signal can be integrated or distributed at different spatial locations. The analysis unit is the core unit of a light scattering system as it converts the scattered signal into surface information using a specific algorithm. A system based on light scattering techniques can be designed to output 'yes or no' results to determine whether the measured surface is defective or not, output a single real number representing surface texture, or be designed with a multilayer machine learning model trained with hundreds of thousands of datasets to measure complex geometrical features.

To go from the surface topography to the light scattering signal is a forward modelling process, while to go from the light scattering signal to the surface information is an inverse problem. The light scattering problem can be highly non-linear and complex. As different surface topographies can have similar scattering patterns, it is sometimes impossible to solve the inverse scattering problem directly (or impossible to obtain a single result). Researchers have developed methods to address the inverse scattering problem, for example the library search method and machine learning methods (see section 6.4). Most methods rely on the generation of a scattering dataset and a specific type of forward scattering model is used to simulate scattering patterns belonging to the type of surface geometry of interest. Different scattering models have been developed with different scattering theories (see section 6.4). Some models are approximation-based, while others are rigorous. Different models also have different computational complexities. Models that have high computational complexity have benefited from the quickly developing modern computational techniques. The following section introduces the basic principles of a light scattering system and is followed by some advanced systems used in manufacturing industry. Advanced algorithms and accelerated computational technologies in recent developments are introduced.

6.2 Basic principles

The basic principle of surface measurement using a light scattering technique is illustrated in figure 6.1. A light source is projected onto a target surface and the scattered light is measured by a sensor. The scattering signal is often recorded by a PC for data processing and the surface information can be obtained through a specific algorithm.

The configuration of the light scattering system can be different depending on different applications, such as measuring surface texture, critical dimensions or defects. The scattered light can be measured by the sensor in an integrated or an angular manner, known as total integrated scattering (TIS) and angle-resolved scattering (ARS), respectively (Stover 2012). ARS is often described by the bidirectional reflectance distribution function (BRDF). TIS and ARS have applications in texture measurement and are described by international standards, such as SEMI MF 1048-1109 (SEMI 2009b) and SEMI ME 1392-1109 (SEMI 2009a), respectively. The schematic diagram of a TIS apparatus and the BRDF concept to describe light scattered from a rough surface are shown in figures 6.2 and 6.3, respectively. Recently, there has been an international round-robin to compare the performance of ARS procedures and methods (Von Finck *et al* 2019). To increase the dynamic range measurement capability of the instrument, the incidence angle and the wavelength of the light source can be adjustable, such as those of angle-resolved scatterometry (Neubert *et al* 1994) and specular spectroscopic scatterometry (Niu *et al* 2001). It is also possible to adjust the polarisation states of the light source, for example, ellipsometers (Liu *et al* 2015).

With the growing demands and fast development of the advanced manufacturing and semiconductor industries, different system designs are needed to meet stringent and task-specific requirements. One example is the extreme ultraviolet (EUV) scatterometer (Orji *et al* 2018). As feature sizes shrink down to several nanometres in the EUV lithography technology, challenges are raised for the development of EUV scatterometers.

Light scattering modelling, as a forward modelling method, is the method to determine scattering patterns given the experimental conditions, such as the surface topography, the type of incident light and the material properties (complex refractive index). A number of computational light scattering models have been

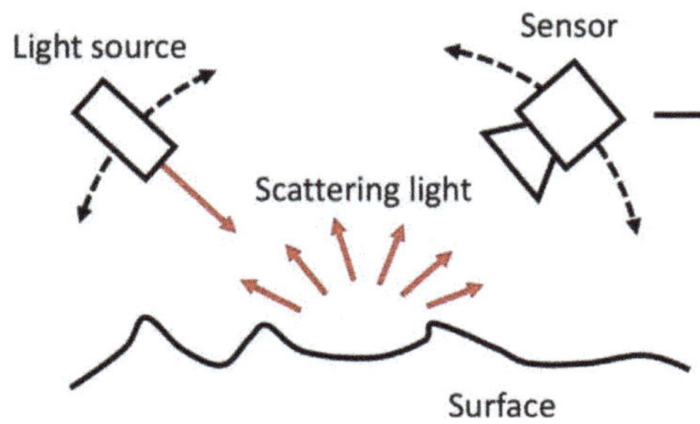

Figure 6.1. Basic principle of surface measurement using the light scattering technique.

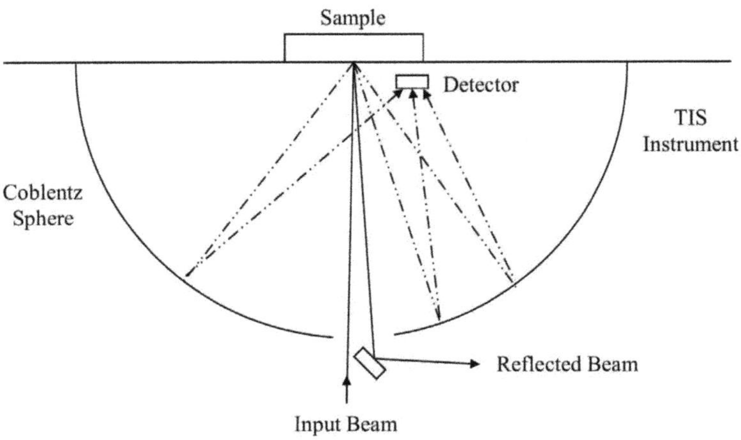

Figure 6.2. Schematic diagram of a TIS apparatus. (Reproduced with permission from Stover (2012). Copyright 2012 SPIE.)

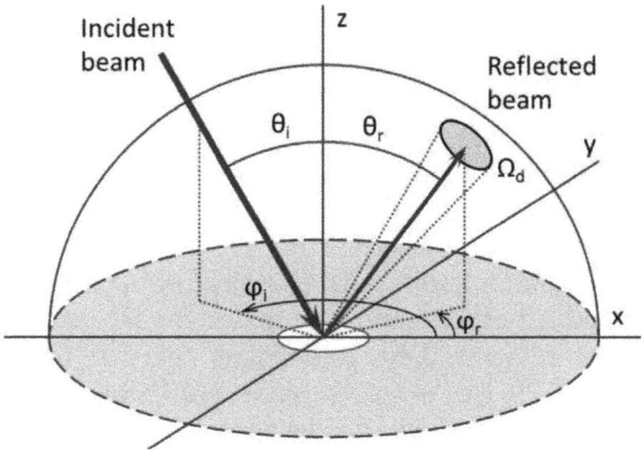

Figure 6.3. Schematic diagram of the BRDF concept to describe light scattered from a rough surface. Ω_d is an element of reflected light, θ_i and φ_i are, respectively, the polar and azimuth angles for the incident light, θ_r and φ_r are, respectively, the polar and azimuth angles for the reflected light (Matsapey *et al* 2013). Copyright IOP Publishing. Reproduced with permission. All rights reserved.

developed according to the different complexities and accuracy requirements of applications. For some applications, where the dimensions of the structures of interest are much larger than the wavelength of the light source, simplified models, such as ray tracing, can be applied (Mouroulis and Macdonald 1997). For more complex cases, where diffraction effects cannot be neglected, more advanced models are required, such as rigorous coupled-wave analysis (RCWA) (Moharam and Gaylord 1981), finite element methods (FEM) (Kato *et al* 2012), finite-difference time-domain (FDTD) methods (Taflove and Hagness 2005), Monte Carlo methods (Delacrétaz *et al* 2012) and boundary element methods (BEMs) (Simonsen 2010), to name just a few. These methods have greatly advanced applications using the light scattering technique and have been widely applied in various industries.

To determine the surface characteristics from the scattering patterns, in contrast to the forward modelling method, is an inverse scattering problem. The inverse scattering problem is difficult to solve directly as the scattering phenomenon is highly non-linear and sometimes there is no unique and closed-form solution. One solution to address the inverse scattering problem is to measure the scattering patterns of surfaces with typical topographies with different parameters and thus to create a database of scattering signals. With the availability of numerical forward models within accuracy requirements, the scattering patterns are often generated via numerical simulation methods rather than from experiments, so as to save time and resources. Once the database is established, any new scattering signals from the experiment can be compared with the signals in the database. A best-fit pattern can be found and thus the characterisation of the surface can be estimated. This is the so-called conventional library search method. The library search method has been used in the scatterometry industry for decades to address the inverse scattering problem (Paz *et al* 2012). The measured scattering signal is often compared with those in the library by minimising the least-squares error. However, the library search method is time consuming, particularly when the library is large. Advanced methods have been developed, such as support vector machine (SVM) methods and optimisation techniques, such as the Levenberg–Marquardt (LM) method (Zhu *et al* 2019), neural network methods (Madsen *et al* 2018) and maximum contributed component regression (MCCR) methods (Zhu *et al* 2017). It should be noted that there is a research trend towards using machine learning methods to address the inverse scattering problem, as such methods are considered as fast and robust (Liu *et al* 2020).

6.3 Advanced systems

6.3.1 Advanced systems for mechanical engineering and manufacturing

A compact light scattering-based instrument has been developed and provided the capability for in-line measurement (Herffurth *et al* 2013). Figure 6.4 shows the implementation of the light-scatter sensor and the scheme of the sensor set-up. The sensor has a drawback in that the area to be evaluated is limited by the spot size of the light source and the area captured by the camera, which has hindered its applications in industry, where often a large area is required for evaluation.

Recently, Herffurth *et al* (2019) combined a light scattering sensor with a robotic arm to perform defect detection on freeform surfaces, which significantly enlarged the measurable area and hence enhanced the applications of the light scattering technique for defect detection in industrial applications. Figure 6.5 shows the implementation of the light scattering sensor combined with an industrial robot during the characterisation of a hyperbolic mirror, and the schema showing the primary sensor components. Figure 6.6 shows the typical characterisation result for a lightweight convex aspheric mirror, where the texture mapping provides information for contamination or defects, such as that at location X4.

Liu *et al* (2019) developed a goniometer-like system which used the light scattering angular spectrum combined with a machine learning method to determine

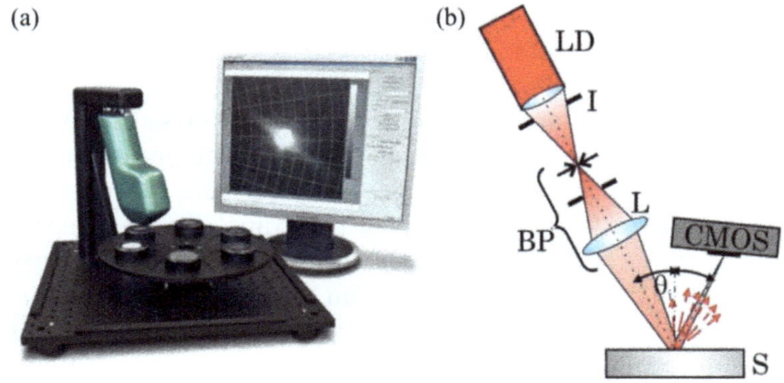

Figure 6.4. (a) Implementation of the light scattering sensor and (b) scheme of the sensor set-up. LD: laser diode; I: iris; L: lens; BP: spatial frequency filter; S: sample. (Reproduced with permission from Herffurth *et al* (2013). Copyright 2013 OSA.)

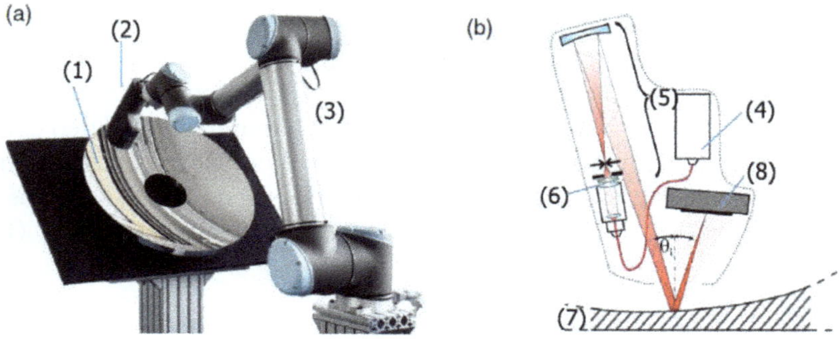

Figure 6.5. Robotic light scattering sensor: (a) photograph of the implementation, (b) scheme of the optical set-up: (1)/(7) sample, (2) sensor head, (3) robot, (4) laser source ($\lambda = 650$ nm) and optical fibre, (5) spatial filter, (6) polariser and (8) CMOS matrix detector. (Reproduced with permission from Herffurth *et al* (2019). Copyright 2019 SPIE.)

defects on microstructured surfaces. The defective and non-defective surfaces were artificially generated and their scattering patterns were simulated using BEM (Thomas *et al* 2020). The surfaces were labelled according to their defective states. The scattering patterns and the labels were used to train a machine learning model. Once the machine learning model was trained, the defective information of the surface can be detected using the measured scattering pattern. Figure 6.7 shows the schema of the proposed method and the experimental set-up. The microstructured surfaces and their measured scattering signals are shown in figure 6.8. With the proposed method, the defective surface could be successfully determined in a fast and robust manner.

The defect detection concept using light scattering and machine learning was further developed and implemented as an on-machine surface defect detection system using a convolutional neural network (CNN)-based deep learning model (Liu *et al* 2020). The system is shown in figure 6.9 and was mounted on a diamond turning machine to perform on-machine experiments. Grating surfaces were

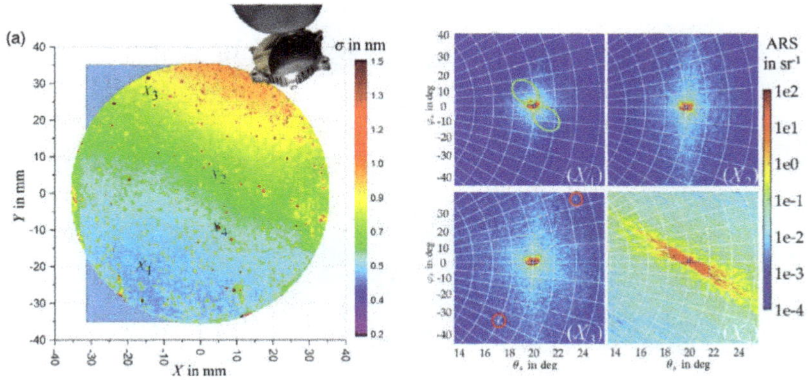

Figure 6.6. Light scattering-based texture analysis of a lightweight convex aspherical mirror. (a) Texture mapping obtained from light scattering, together with a photograph of the sample. (X_1) to (X_4): light scattering distribution recorded at four different positions on the sample. (Reproduced with permission from Herffurth *et al* (2019). Copyright 2019 SPIE.)

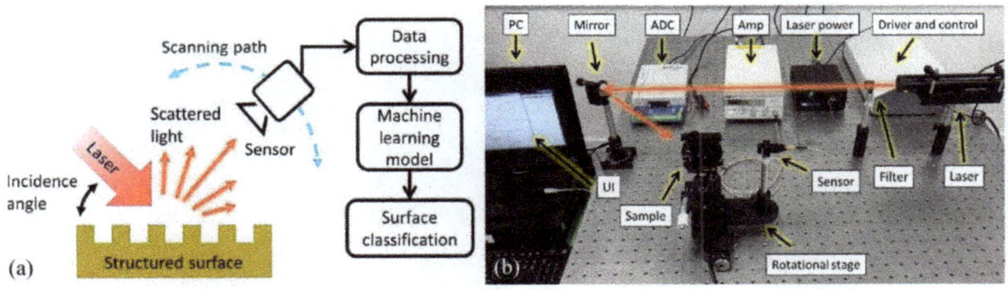

Figure 6.7. Light scattering method for defect detection: (a) schema of the proposed method and (b) the experimental set-up (Liu *et al* 2019).

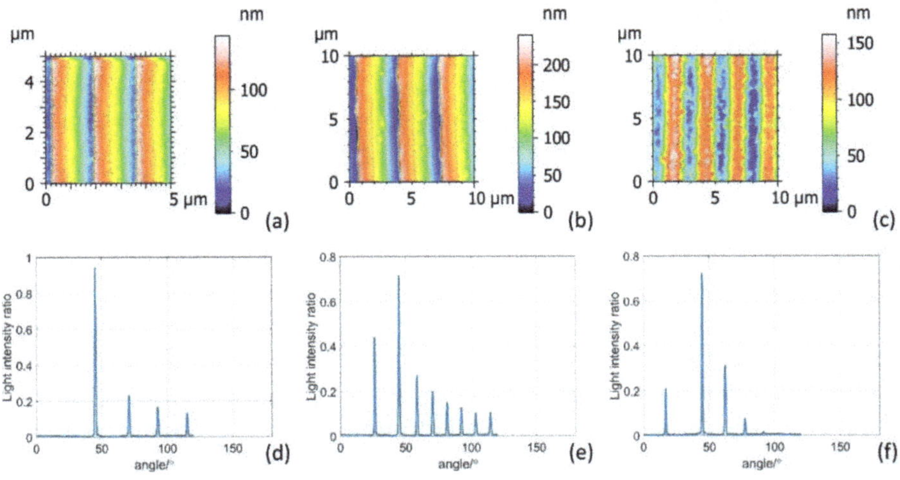

Figure 6.8. Microstructured surfaces and their measured scattering signals, (a)–(c) atomic force microscopy measurement results and (d)–(f) measured scattering signals (Liu *et al* 2019).

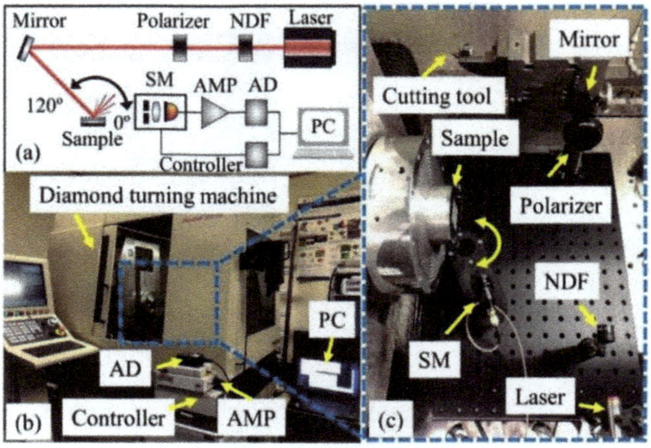

Figure 6.9. Prototype system of the on-machine surface defect detection system: (a) diagram, (b) set-up for the on-machine experiment and (c) enlarged view for the system (Liu *et al* 2020).

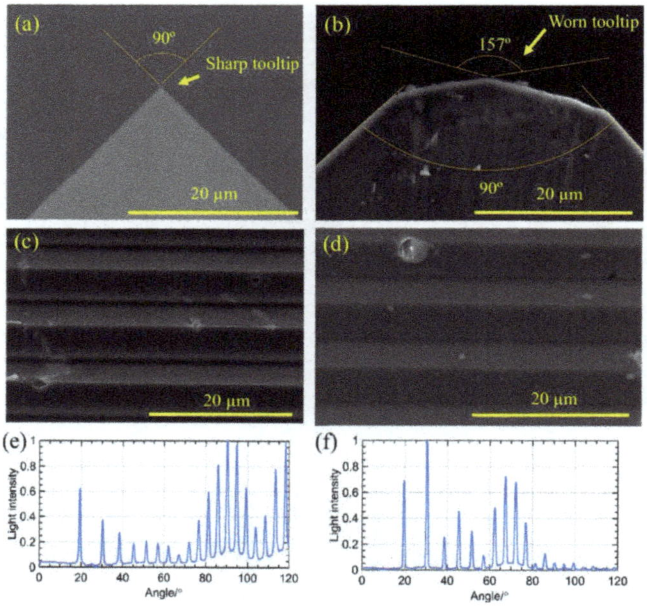

Figure 6.10. Experiment results demonstrated two microstructured surfaces machined by two diamond tools, i.e. a sharp tool and a worn tool: (a) and (b) SEM images of the tools, (c) and (d) SEM images of the machined surfaces, and (e) and (f) measured scattering signals (Liu *et al* 2020).

demonstrated and machined by sharp and worn diamond tools. Figure 6.10 shows one set of experimental results demonstrating two microstructured surfaces machined by two diamond tools, i.e. a sharp tool and a worn tool, representing a non-defective and defective surface, respectively. Using the measured scattering signals, shown in figure 6.10, the defective surface machined with the worn tool could be successfully determined.

Imlau *et al* (2016) developed a sensor based on a light scattering technique, named the 'riblet sensor', to evaluate the coatings of microstructured surfaces which were designed to reduce air resistance for aeroplanes. Figure 6.11 shows the schema and photograph of the experimental set-up.

As a demonstration of the effectiveness of the riblet sensor, degraded and non-degraded riblet surfaces were tested. Figure 6.12 shows scanning electron microscopy (SEM) results for non-degraded and degraded riblet structures and figure 6.13 shows the resulting scattered intensities. The scattering patterns showed significant differences and hence they could be used to evaluate the quality of the surfaces.

Schröder *et al* (2014) developed an instrument which was able to perform spectral angle-resolved scattering measurements with a wide spectral range (250–1500 nm) for high-quality optical components. The schema of the instrument is shown in figure 6.14.

The advantage of the spectral angle-resolved light scattering instrument is that it uses a solid-state laser pumped optical parametric oscillator light source that can continuously output light over the spectral range with a spectral bandwidth smaller than 0.05 nm. The scatterometer was an improvement of the instrument for 3D ARS

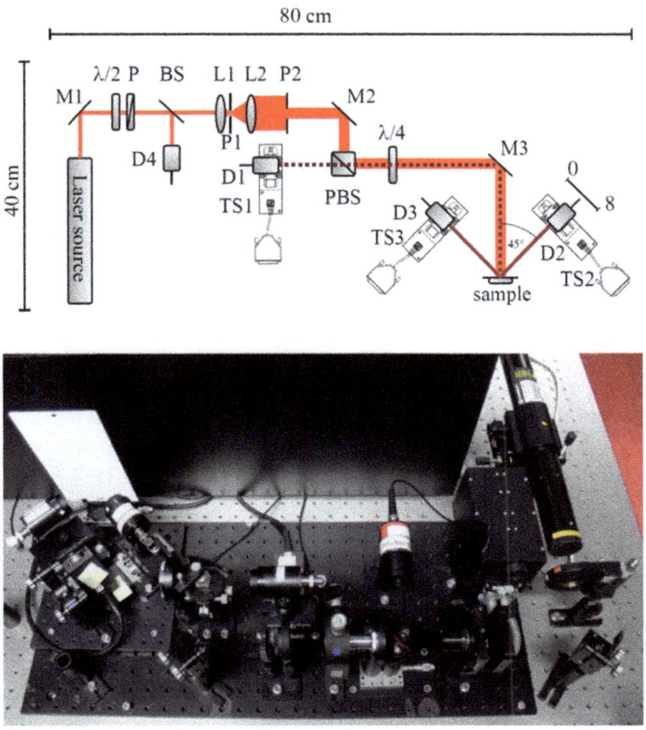

Figure 6.11. Schema and photograph of the experimental set-up allowing for the determination of the angular intensity distribution of waves scattered from a riblet sample. D1–4: SI-PIN photodiodes; TS1–3: mounted of linear translation stages; **BS**: beam splitter; $\lambda/2$: half-wave plate; P: polariser; M1–3: mirrors; L1–2: lenses; P1–2: pinholes; **PBS**: polarising beam splitter cube; $\lambda/4$: quarter-wave plate. The 0 mm and 8 mm positions are marked for TS2 as examples. (Reproduced with permission from Imlau *et al* (2016).)

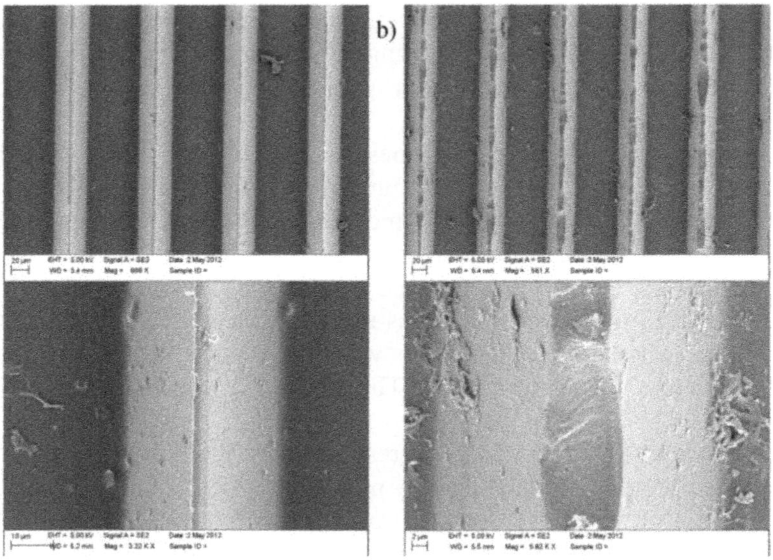

Figure 6.12. SEM data for (a) a non-degraded and (b) a degraded riblet structure. (Reproduced with permission from Imlau *et al* (2016).)

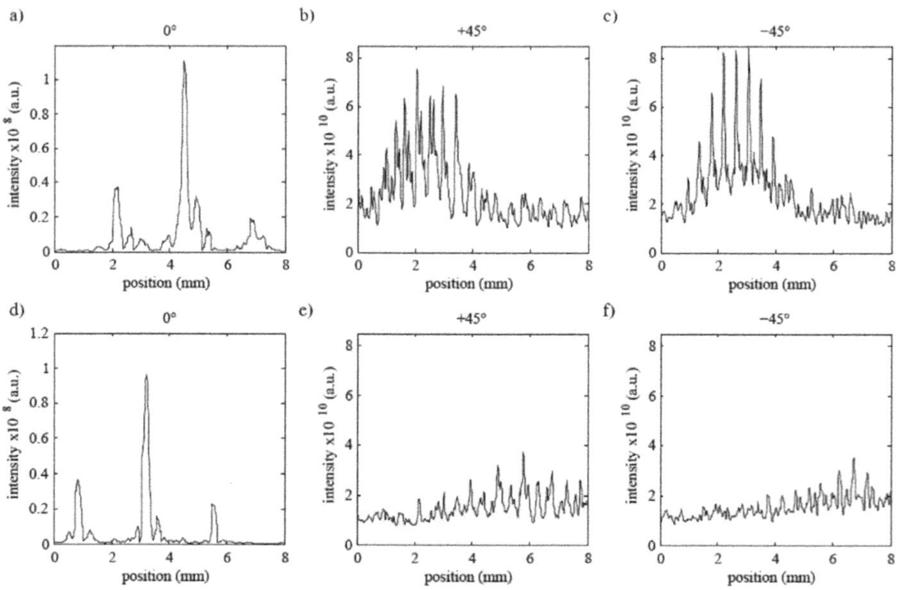

Figure 6.13. Scattered intensity for the 0° and ±45° patterns as a function of position for (a)–(c) a non-degraded and (d)–(f) a degraded riblet structure. (Reproduced with permission from Imlau *et al* (2016).)

measurements—ALBATROSS, developed by Schröder *et al* (2011)—but equipped with a broadband beam preparation and detection system. In addition to the sensitivity, the accuracy, reproducibility and linearity were improved. A schema and photograph of the ALBATROSS are shown in figure 6.15.

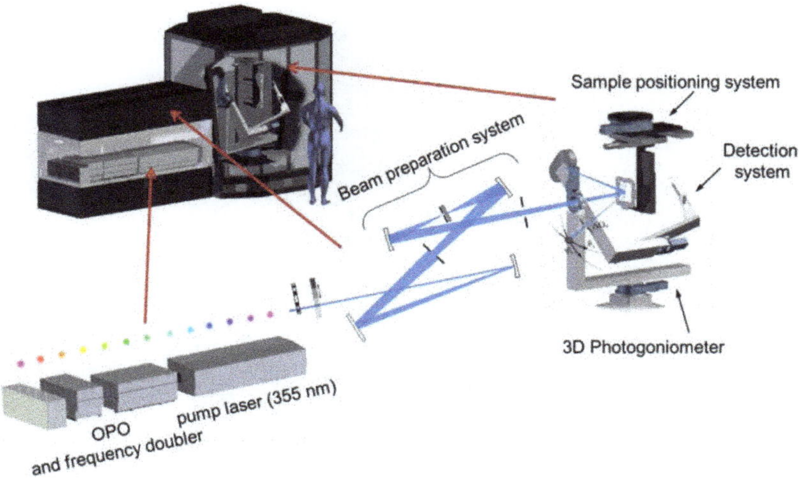

Figure 6.14. Schema of an instrument for spectral angle-resolved light scattering measurements. (Reproduced with permission from Schröder *et al* (2014). Copyright 2014 OSA).

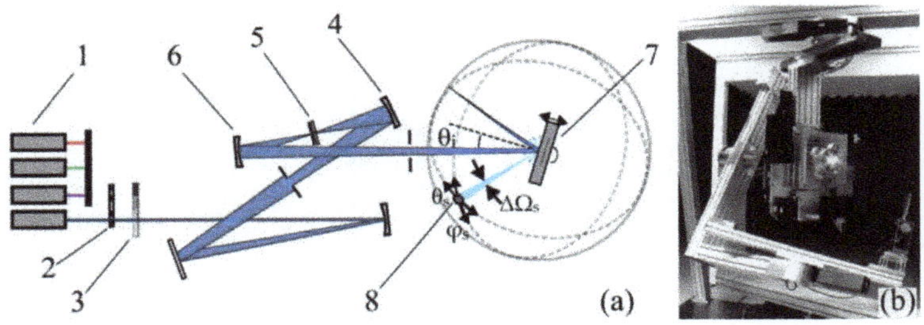

Figure 6.15. The ALBATROSS instrument for angle-resolved scattering measurement in the UV–VIS–IR spectral range. (a) Schema. (b) Photograph showing a sample (centre) mounted onto the sample positioning system, as well as the detector and 3D goniometer. 1: multiple laser sources with different wavelengths; 2: chopper; 3: neutral density filter; 4: spherical mirror; 5: pinhole; 6: spherical mirror; 7: sample; 8: detector aperture. (Reproduced with permission from Schröder *et al* (2011). Copyright 2011 OSA.)

6.3.2 Advanced systems for semiconductor manufacturing

Semiconductor technology has been rapidly developed over the last decades, aiming to manufacture extremely small 3D structures with nanoscale features. Currently, EUV lithography is the dominant technology for the next generation of semiconductor manufacturing of nanoscale structures, so to reduce the wavelength of the light source it is essential to reduce the size of machined structures. Precision nanoscale structures also raise challenges for conventional measurement techniques. One solution is to develop EUV-based scattering techniques.

The Physikalisch-Technische Bundesanstalt (PTB) has developed an EUV scatterometer using the EUV reflectometry facility at the electron storage ring BESSY II (Klein *et al* 2006, Scholze *et al* 2006, Kato and Scholze 2010, Scholze *et al* 2011). Figure 6.16 shows the mechanics of the PTB EUV reflectometer. Their soft x-ray radiometry beamline was used with a plane grating monochromator which

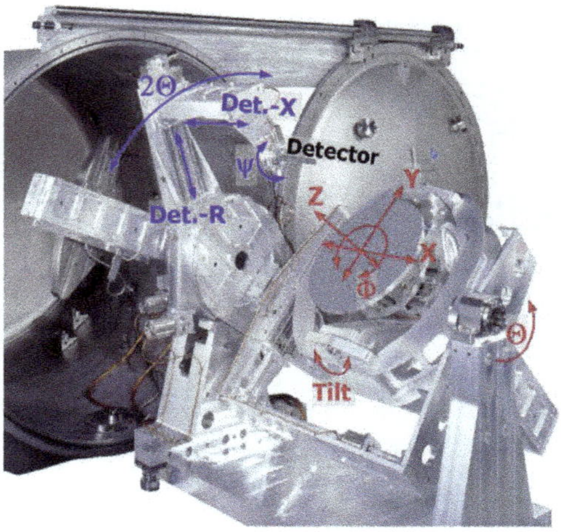

Figure 6.16. Mechanics of the EUV reflectometer. (Reproduced with permission from Scholze *et al* (2006). Copyright 2006 SPIE.)

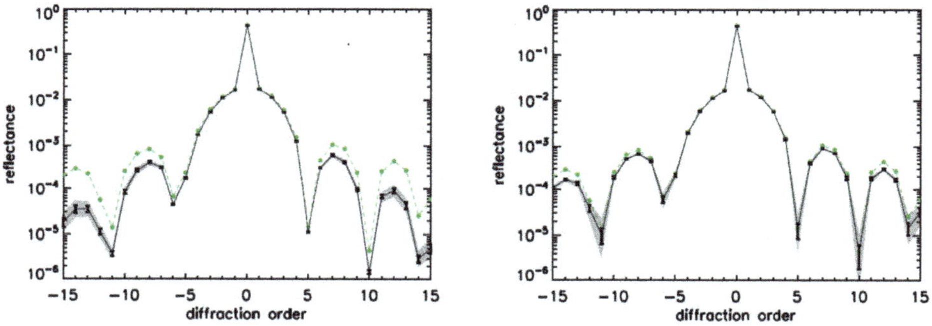

Figure 6.17. Result of a Monte Carlo simulation of LER (left) and LWR (right). Grey: range of reflectance at an error of 10% of CD. Black: mean and standard deviation for each diffraction order. Green: Fraunhofer diffraction pattern of the undisturbed periodic grating. (Reproduced with permission from Kato and Scholze (2010). Copyright 2010 SPIE.)

covered the spectral range from 0.7 to 35 nm and was designed to provide highly collimated radiation (Scholze *et al* 2001). The short wavelength of EUV was advantageous since it provides more propagating diffraction orders compared to longer wavelength radiation. The short wavelength also increases the sensitivity to small structural features, particularly high spatial frequency texture.

The PTB EUV scatterometer was applied for the measurement of absorber lines with a trapezoidal cross-section on semiconductor photomasks and of ultra-smooth surfaces of EUV multilayer mirrors. The results showed that structured texture significantly affected the scattered diffraction intensities which must be included in the reconstruction algorithms using inverse modelling by FEM. The result of a Monte Carlo simulation of line edge roughness (LER) and line width roughness (LWR) is shown in figure 6.17.

A coherent EUV scatterometry microscope (CSM) to measure EUV patterns based on the coherent-diffraction-imaging method has been developed (Harada *et al* 2011, Harada *et al* 2013, Fujino *et al* 2015). A schema of the CSM is shown in figure 6.18. Rather than capturing the scattering patterns by scanning in an arc, the diffraction patterns were captured by a camera.

Diffraction images from EUV masks with no defects and with different types of defects are shown in figure 6.19, including oversize defects, undersize defects and bridge defects. As different types of defects have different diffraction patterns, the diffraction patterns can be used for defect detection with the EUV masks.

Gross *et al* (2009) evaluated the measurement uncertainties in EUV scatterometry. The experiments were carried out using the EUV reflectometer shown in figure 6.20. The scheme of an EUV line-space structure is shown in figure 6.21. The influence of certain presumed and fixed model parameters, i.e. the thicknesses of the two capping layers and the four widths in the periodically repeated groups of the multilayer system were analysed. It was found that the impact of model uncertainties was critical and led to uncertainties of up to 3% for all the parameters investigated, which was comparable to the detector-noise related uncertainties. The model uncertainties induced systematic shifts of the results for the capping layer thickness and sidewall angle. The CD for the bottom and top of the line-space structure also had significantly increased variations. The standard deviation of the sidewall angle was greater than 1.5°. It was observed that the height of the line-space structure and its mean CD were relatively stable with respect to the studied model-based uncertainties. Furthermore, experiments revealed a strong correlation between the thicknesses of the capping layers and the sidewall angle.

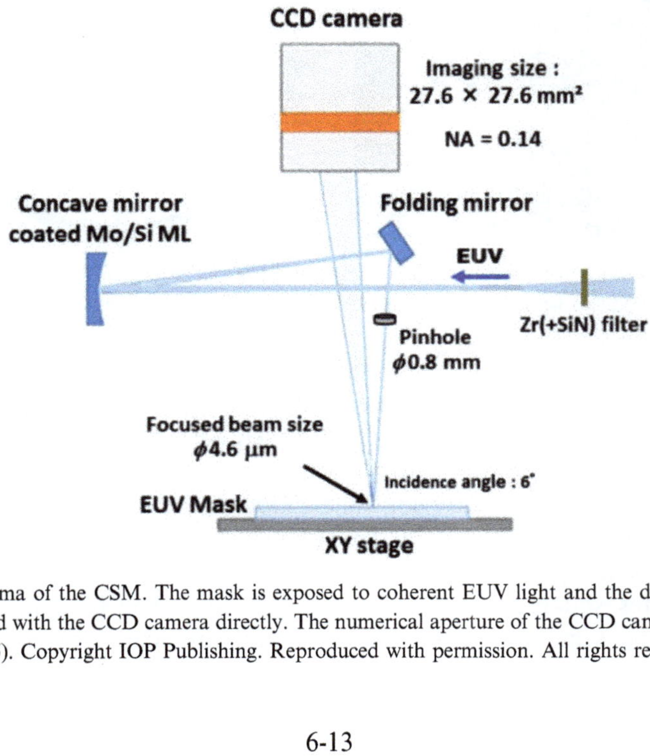

Figure 6.18. Schema of the CSM. The mask is exposed to coherent EUV light and the diffraction from the pattern is recorded with the CCD camera directly. The numerical aperture of the CCD camera is around 0.14 (Fujino *et al* 2015).

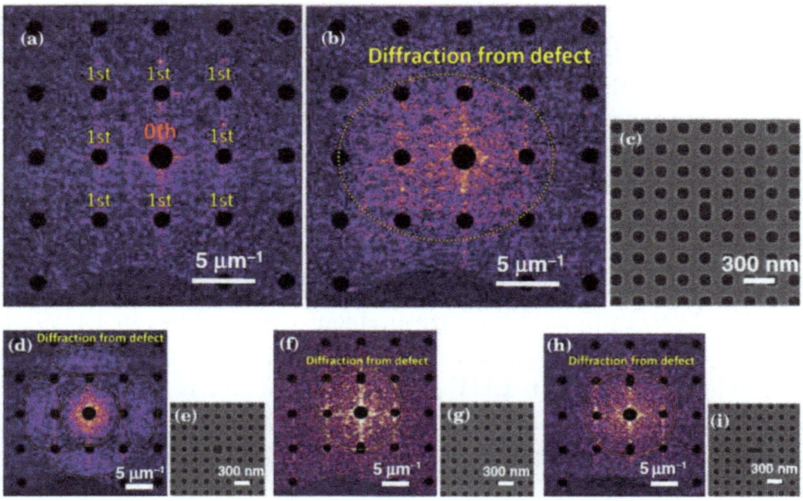

Figure 6.19. Diffraction image from an EUV mask with a 112 nm hole pattern and SEM images. CSM result of (a) no defect region, (b) line-end 76 nm oversize defect, (d) 38 nm oversize defect, (f) 20 nm undersize defect and (h) 80 nm bridge defect. (c), (e), (g), and (i) SEM images of each defect (Fujino *et al* 2015). Copyright IOP Publishing. Reproduced with permission. All rights reserved.

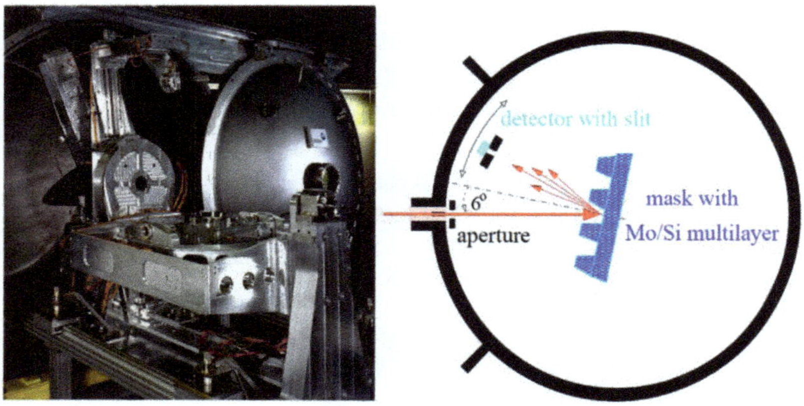

Figure 6.20. Spectroscopic reflectometer operating in the EUV range (0.7–35 nm) and the scheme of the measurement. (Reproduced with permission from Gross *et al* (2009). Copyright 2009 SPIE.)

Most scatterometry techniques measure the averaged signal over an area equal to the spot size of the light source. To measure local topography parameters, conventional split-type scatterometry can be integrated into a microscope. Madsen and Hansen (2016a) presented an imaging scatterometry technique, where only a small portion of the surface was examined, which made use of the limited field of view and numerical aperture of the objective. The imaging-based scatterometry was capable of measuring local topographic parameters of gratings spanning an area down to a few micrometres squared with nanometre accuracy. The imaging scatterometer could easily find areas of interest on the centimetre scale and measure multiple

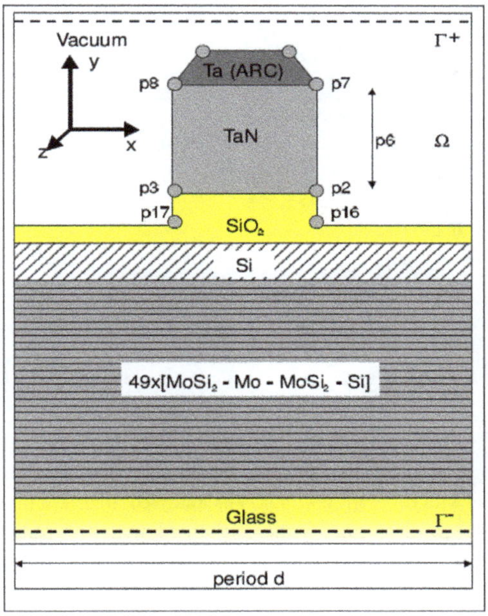

Figure 6.21. Scheme of an EUV line-space structure composed of three trapezoidal absorber layers of different materials. (Reproduced with permission from Gross *et al* (2009). Copyright 2009 SPIE.)

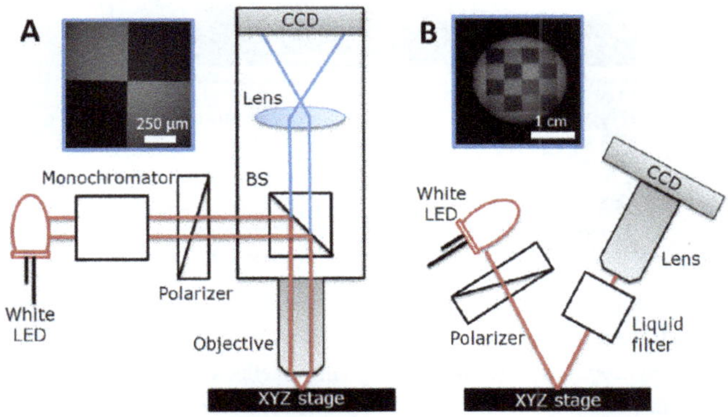

Figure 6.22. Experimental set-ups for imaging scatterometry. (A) Sketch of the system built into an optical microscope. The insert shows the intersection of four fields of a multi-structured sample. (B) Schema of a split configuration system with filtering on the output side. The insert shows sixteen fields of a multi-structured sample (Madsen and Hansen 2016a, copyright OSA).

segments simultaneously. Two types of scatterometers, one built into an optical microscope and one in a split configuration, were demonstrated in the experiments, as shown in figure 6.22.

The results for the two systems with an optical microscope and split configuration are shown in figures 6.23 and 6.24, respectively. The imaging scatterometers had an analysed area of (12 × 12) μm whilst that with the split-type scatterometer was (300 × 300) μm, which could be selected to fit the requirements of different applications.

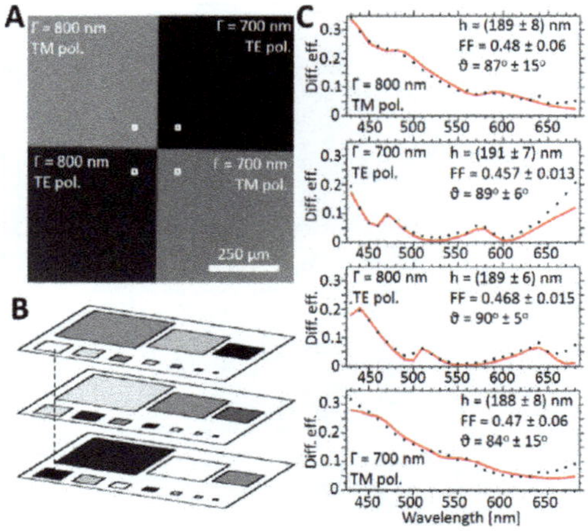

Figure 6.23. Imaging scatterometry data obtained in the microscope configuration. (A) Pixel based diffraction efficiency image of four areas with different gratings. The analysed areas are marked with white squares and have the size of (12 × 12) μm. (B) Illustration of the pixel based multi-dimensional analysis of the diffraction efficiency images. (C) Wavelength dependent diffraction efficiencies and best fit for the pixels indicated in (A). The ± denotes the 95% confidence limits of the fitted parameters (Madsen and Hansen 2016a, copyright OSA).

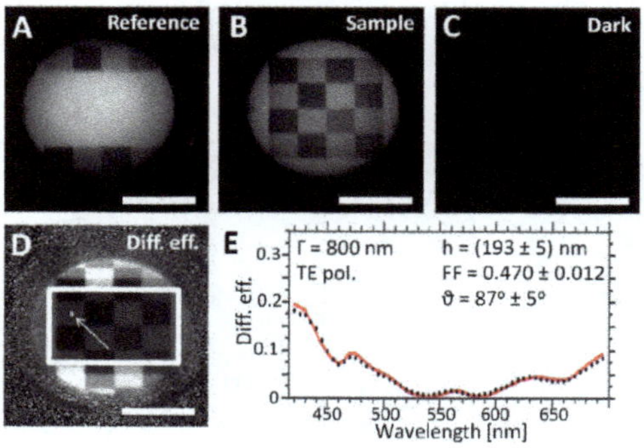

Figure 6.24. Imaging scatterometry with a split configuration scatterometer. (A) Reference, (B) sample and (C) dark images obtained with the same settings of the camera. (D) Computed diffraction efficiency for each pixel. The pixels have been binned (2 × 2) for noise reduction giving an effective pixel size of (300 × 300) μm. Only the diffraction efficiencies of the pixels within the white box are valid, as a reference sample with structures outside this area has been used. (E) Wavelength dependent diffraction efficiencies and best fit for the area with a pitch of 800 nm and grating orientation parallel to the polarisation of the light. The analysed pixel is indicated in (D) with a very small white rectangle. Scale bar in (A–D) is 1 cm (Madsen and Hansen 2016a, copyright OSA).

Madsen *et al* (2015) integrated a fibre-based spectrometer into a light microscope to characterise nano-textured surfaces. The adapted microscope had two detectors, a CCD camera used to easily find an area of interest and a spectrometer for the spectral measurement. The schema of the experimental set-up is shown in figure 6.25. It was demonstrated that the microscope had a resolution in the nanometre range for measurement of the topographic parameters—height, width and sidewall angle of a periodic grating—even in an environment with high levels of vibration, such as a production facility with heavy manufacturing equipment. The measured surface

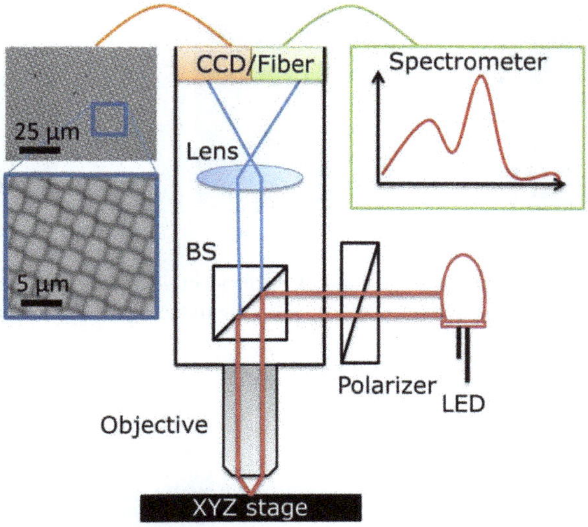

Figure 6.25. Schema of experimental set-up using scatterometry integrated into a commercial light microscope (Madsen *et al* 2015, copyright OSA).

could be translated during acquisition, as long as the beam spot was kept inside an area with homogeneous structures, which made the proposed microscope well suited for implementation in a production environment.

6.4 Advanced scattering algorithms

The development of light scattering techniques does not only require sophisticated hardware design but also requires advanced algorithms to support the forward modelling and solve the inverse scattering problem. As discussed, a number of forward models have been developed, such as ray tracing models, RCWA, FEM, FDTD, Monte Carlo methods and BEM. Simplified models, such as ray tracing, may not be sufficient when diffraction effects cannot be ignored, such as with surfaces with micro-/nano-structures. RCWA is most efficient with periodic structures and cannot be applied for randomly rough surfaces. FEM and FDTD methods are often used to model scattering fields in a small region as they have very high computational complexity.

An accurate and efficient numerical approach to simulate electromagnetic wave scattering from 3D surfaces was proposed by Simonsen *et al* (2010). The method makes use of the Müller equations (Müller 2013) and an impedance boundary condition for a 3D rough surface, which yields a pair of coupled 2D integral equations for the sources on the surface in terms of which the scattered field is expressed through the Franz formulae (Franz 1948). Through this approach, the full angular intensity distribution of the scattered field can be calculated. Figure 6.26 shows one example of the full angular intensity distribution of the scattered field for a surface.

Coupland and Nikolaev (2019) proposed a boundary source method (BSM) applied to the vector calculation of electromagnetic fields from a 3D surface defined

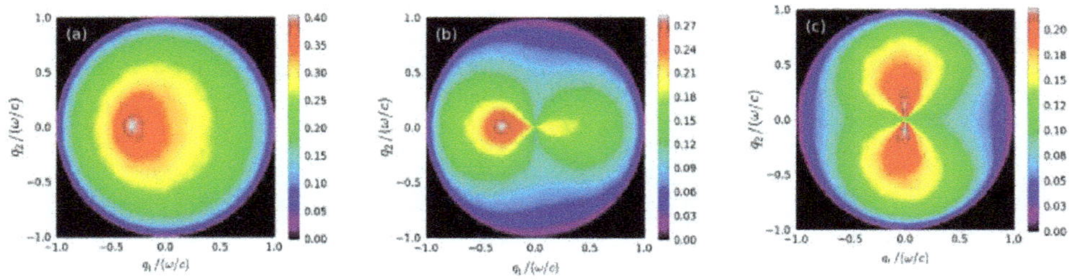

Figure 6.26. The full angular intensity distribution of the incident *p*-polarised light scattered into (a) *p*- and *s*-polarised light, (b) *p* polarisation and (c) *s* polarisation. (Reproduced with permission from Simonsen *et al* (2010). Copyright 2010 APS.)

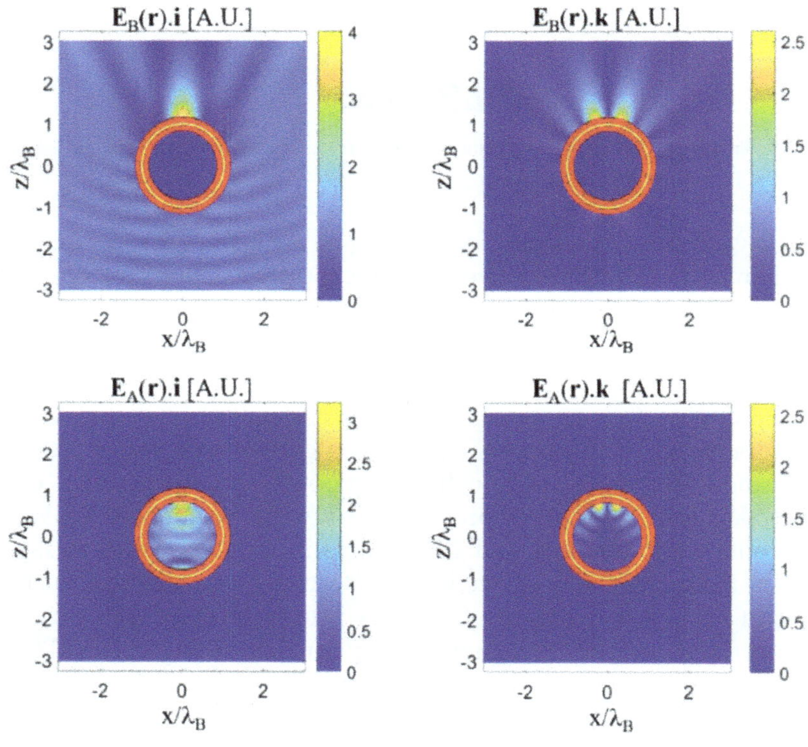

Figure 6.27. Extinction of the sum of the scattered field and incident field inside of the sphere and the extinction of the transmitted field outside the sphere. E_A and E_B are the scattering fields in medium A and B, respectively; *i* and *k* are scattering fields along the axes *x* and *z*, respectively; [A.U.] denotes that the result is shown as a function of the wavelength (Coupland and Nikolaev 2019, copyright OSA).

by the interface between homogeneous, isotropic media. The BSM method provides an accurate solution to electromagnetic scattering from a band-limited surface. Figure 6.27 shows a modelling result for the scattering from a sphere. Figure 6.28 shows the comparison results of the scattering fields with the Mie series (Hergert and Wriedt 2012) and the results show good agreement.

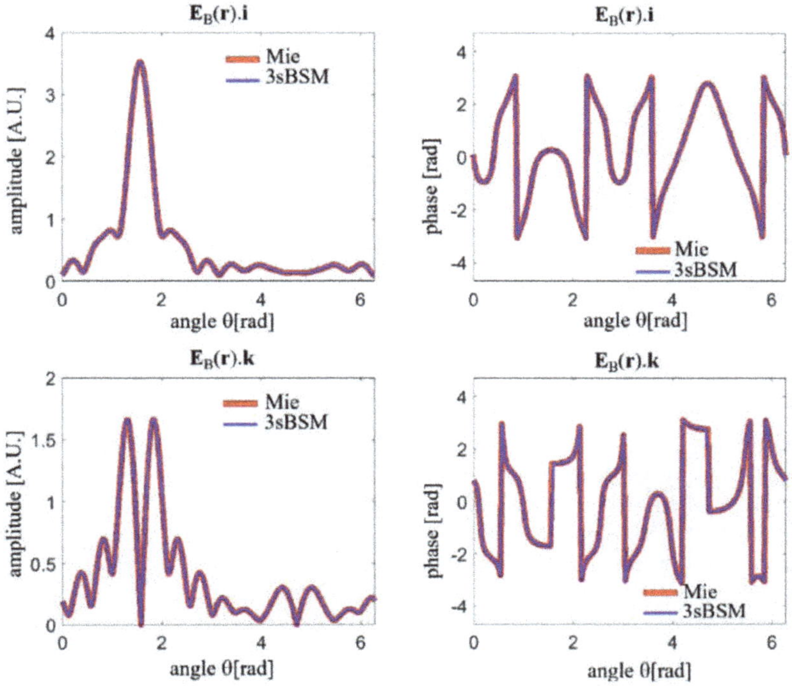

Figure 6.28. Comparison of the scattered fields with Mie series. E_B is the scattering field in medium B, i.e. outside the sphere; i and k are the scattering fields along the axes x and z, respectively; [A.U.] denotes that the result is shown as a function of the wavelength (Coupland and Nikolaev 2019, copyright OSA).

The library search method is a conventional method used in scatterometry to solve the inverse scattering problem (Madsen and Hansen 2016b). However, when the range and resolution of the measurement increase, the library becomes large and the computational speed becomes slow. The target of library searching is to minimise the least-squares function (LSQ), but this process cannot correct systematic errors, which are often caused by incorrect assumptions about the corresponding measurement accuracies of the scattering efficiencies. Henn *et al* (2012) developed a maximum likelihood estimation (MLE) method to determine the statistical error model parameters from measurement data. By using simulation data, the results showed that the MLE method was able to correct the systematic errors present in LSQ results and improved the accuracy of scatterometry. Figure 6.29 shows the comparison results for reconstructed sidewall angles with approximately 95% confidence intervals for measured data from EUV scatterometry.

Figure 6.30 shows the comparison results for reconstructed sidewall angles with approximately 95% confidence intervals for measured data from a DUV scatterometer using LSQ and MLE methods. Note that the mean sidewall angle obtained by SEM was approximately 86°, which demonstrates that the MLE method has better accuracy than the LSQ method.

To improve the computational speed issue for the library search method, machine learning methods based on neural networks have been developed. Neural networks,

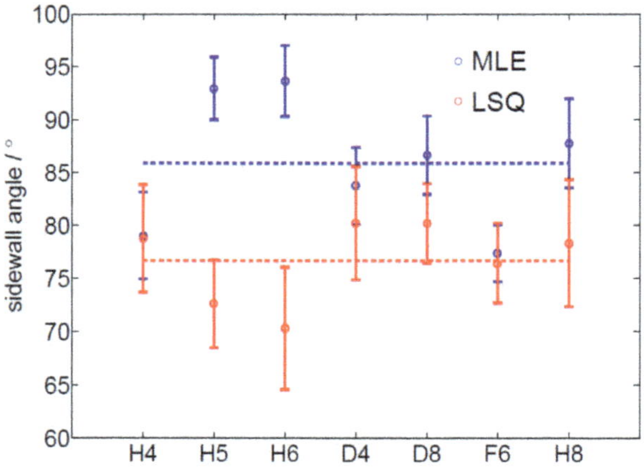

Figure 6.29. Reconstructed sidewall angles with approximately 95% confidence intervals for measured data from EUV scatterometry, the dotted lines represent the mean values for the two methods. H4, H5, H6, D4, D8, F6 and H8 are different designs of the EUV mask. H4: 840 nm period, 140 nm botCD; H5: 420 nm period, 140 nm botCD; H6: 280 nm period, 140 nm botCD; D4, D8, F6 and H8: 720 nm period, 540 nm botCD. botCD: bottom CD of the photomask (Henn *et al* 2012, copyright OSA).

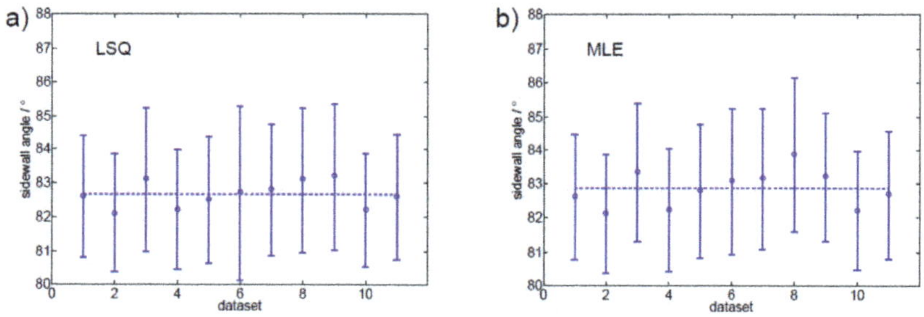

Figure 6.30. Reconstructed sidewall angles with approximately 95% confidence intervals for measured data from the DUV scatterometer (a) for LSQ and (b) for MLE, the dotted lines represent the mean values (Henn *et al* 2012, copyright OSA).

as demonstrated by Madsen *et al* (2018), are often densely connected as shown in figure 6.31. A neural network was used for defect detection and the defect was introduced as a sinusoidal variation of the grating interface. The scattering properties of the gratings were modelled using the RCWA method, and defects were estimated with the neural network. The input layer had a node for each wavelength simulated. The hidden layer had a total of ten nodes, and the output layer had a single node finding the defect magnitude. The nodes from the input layer were connected to the hidden layer through a weighted Tansig transfer function (Ramakrishnan *et al* 2008). In the same manner, all nodes in the hidden layer were connected to the output node through a Purelin transfer function (Hmamouchi *et al* 2016). The method made it possible to avoid the time-consuming library generation/search strategy. It was found that defects were accurately detected and the network was demonstrated to be faster and more versatile than a library search for related structures.

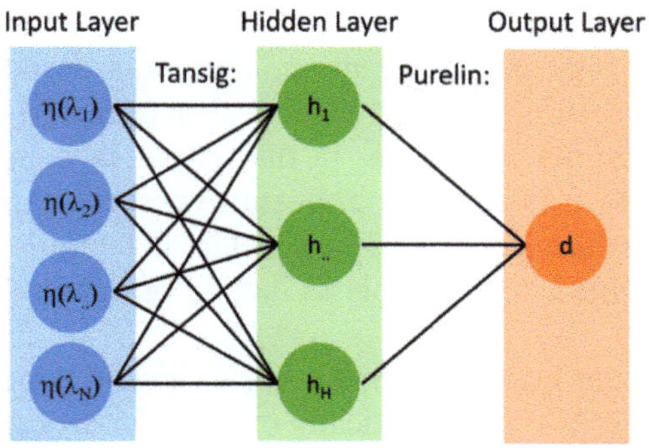

Figure 6.31. Sketch of the neural network (Madsen *et al* 2018, copyright OSA). The input layer has a node for each wavelength simulated. The hidden layer has a total of ten nodes and the output layer has a single node finding the defect magnitude *d*. The nodes from the input layer are connected to the hidden layer through a weighted Tansig transfer function. In the same manner, all nodes in the hidden layer are connected to the output node through a Purelin transfer function. $\eta(\lambda_{1...N})$ are 121 nodes in the input layer for each wavelength, h_{1-H} are nodes in the hidden layer, and *d* is the single node finding the defect magnitude.

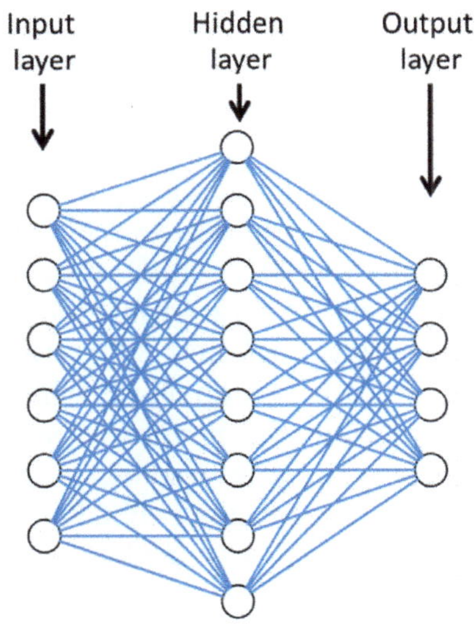

Figure 6.32. Example neural network with three layers: input layer, hidden layer and output layer. The (simplified) example shows six input neurons (corresponding to a scattering spectrum divided into six bins) and four output neurons (corresponding to four classes—either different surface types or defective states, or both) (Liu *et al* 2019).

Liu *et al* (2019) developed a machine learning method to extract the defect information for microstructured surfaces from light scattering signals. Figure 6.32 shows the neural network with three layers, i.e. an input layer, a hidden layer and an output layer. The neurons in the input layer represented the intensities of the far-field scattering signal contained within each angular bin. The classifier interpreted surface types and states (i.e. defective, non-defective) as classes; each class was associated with a neuron containing a real 0–1 value. After the network was trained, the output numbers represented the probability that the input spectrum corresponds to that class. The number of neurons in the input layer was designed according to the number of bins of the scattering signal, which was determined by the scanning range and resolution. The number of neurons in the output layer was determined by the number of considered surface types and sub-types corresponding to one non-defective and multiple defective states. In a defect detection system dedicated to a single surface type, it could be as simple as having only one defective type and one good type (two classes).

A deep learning model was developed in an on-machine defect detection system which made use of a CNN to automatically extract features from scattering signals (Liu *et al* 2020). The CNN is shown in figure 6.33. The CNN was designed as four convolutional layers, two pooling layers and two fully connected layers. The input of the neural network was the scattered signal captured by the sensor over a range of angles. The output of the CNN was two real numbers representing the probabilities of non-defective and defective states for the measured surface. It was shown that combining light scattering with CNN-based deep learning benefited from automatic feature learning and extraction, which made the proposed method capable of learning the complex patterns from the scattering signals. Comparisons with the SVM and library search methods also demonstrated the advantages of the proposed CNN-based method in terms of accuracy and computational time.

Other methods have also been introduced to improve the computational speed of the library search method. A heuristic search algorithm and a robust correction method were proposed to realise a fast library search and to achieve more accurate results for scatterometry (Zhu *et al* 2015). Instead of searching in the signature library, an extra constructed Jacobian library using the principle of gradient-based iteration algorithms (Prudêncio and Ludermir 2003) was designed to perform the search procedure, by which a fast search could be achieved for an arbitrary scale library. A robust correction procedure was performed based on the searched optimal parameter set to obtain more accurate results. A model-free method, called maximum contributed component regression (MCCR), was proposed by Zhu *et al* (2017). Canonical correlation analysis was used to estimate the maximum contributed components from the pairwise relationship of unlabelled data with labelled

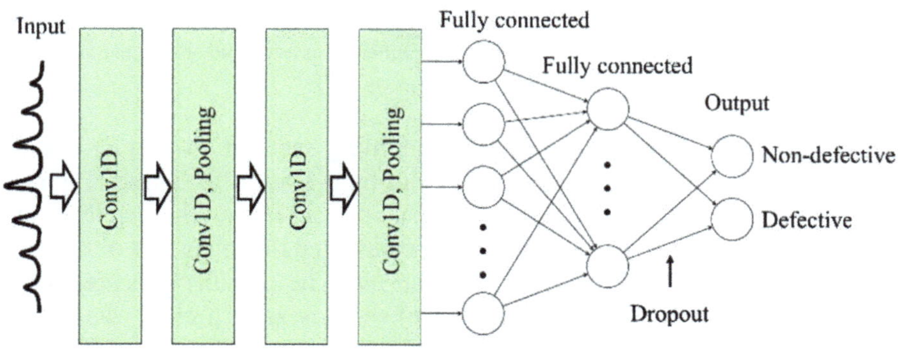

Figure 6.33. Designed CNN (Liu *et al* 2020, copyright OSA).

data. The maximum contributed components were used to guide the solution of the inverse problem based on conventional regression methods. Experimental results on both synthetic and real-world semiconductor datasets demonstrated the effectiveness of the proposed method given a small amount of labelled data. Figure 6.34 shows the comparison of the MCCR method to principal component regression (PCR) and a partial-least-squares method (PLS). R-squared (RSQ) (Cameron and Windmeijer 1997) was used to evaluate the performance of the proposed method. The results show that the MCCR method has better performance than both PCR and PLS methods.

A method combining an SVM and LM algorithm for fast and accurate parameter extraction has been proposed (Zhu *et al* 2019). The SVM technique was introduced to pick out a sub-range of the parameters, in which an arbitrary selected initial solution for the LM algorithm was then used to achieve the global solution with higher accuracy and faster speed. Simulations and experiments conducted on a 1D silicon grating and a deep-etched multilayer grating demonstrated the feasibility and efficiency of the proposed method. Figure 6.35 shows comparison results of the proposed SVM/LM algorithm and those when using solely an LM algorithm.

6.5 Accelerated computational technologies

Light scattering modelling methods, particularly models to calculate the scattered field from a 3D surface, often have high computational complexity. Also, the inverse scattering problem often relies on a large dataset to find the best solution, where the dataset is usually generated by computational simulations, which can also be time consuming. The solutions to solve the inverse scattering problem based on optimisation methods and machine learning methods are based on iteration routines which also have high computational complexity. Most of these processes require numerous computational resources. The development of advanced parallel computation techniques, such as high-performance computing (HPC) and graphics

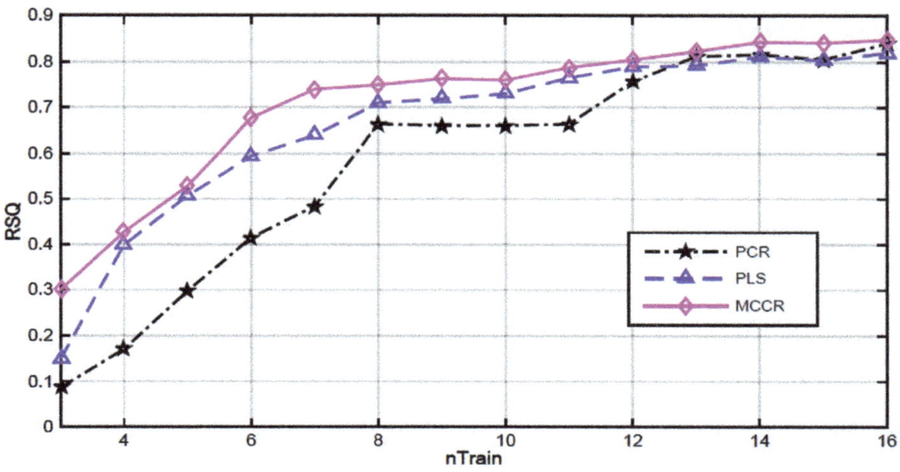

Figure 6.34. Comparison of our MCCR method to PCR and PLS (Zhu *et al* 2017, copyright OSA).

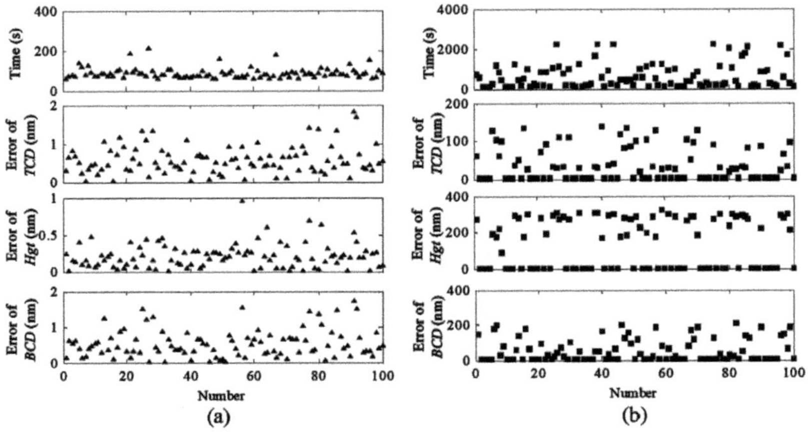

Figure 6.35. Time consumptions and extracted errors of the 1D silicon grating by (a) the proposed SVM/LM algorithm and (b) the LM algorithm, respectively. The subplots from top to bottom correspond to the time consumption, extracted errors of TCD, Hgt and BCD, respectively. TCD: top CD; Hgt: height; BCD: bottom CD (Reproduced with permission from Zhu *et al* (2019).)

processor units (GPUs) has played an important role to address these computational challenges.

Liu *et al* (2020) used an HPC (the University of Nottingham's Augusta HPC service) to generate the scattering patterns for tens of thousands datasets in a parallel manner. The process may take weeks to generate the datasets in a modern high-performance PC and was reduced to hours using the HPC. The HPC had 115 compute nodes with forty processors and 192 GB RAM for each node. A GPU (NVIDIA Quadro P5000) was also used to train machine learning (Liu *et al* 2019) and specifically deep learning (Liu *et al* 2020) models. The GPU had 2560 CUDA cores and 16 GB RAM. Using the GPU, the time for training was reduced from hours to minutes. Using both (HPC and GPU) approaches, the time-consuming simulation and training process was improved to be more practical.

Hidayetoglu *et al* (2017) and Hidayetoglu *et al* (2018) used NCSA Blue Waters, supercomputers with 4224 GPU nodes to solve an inverse scattering problem. The distorted-Born iterative method (DBIM) was employed as an iterative solver. In each iteration, the required forward problems were distributed among computing nodes equipped with GPUs and solved with a multi-level fast multipole algorithm. Tomographic reconstruction of a synthetic object with a linear dimension of 100 wavelengths was obtained on 256 GPUs. The results showed that DBIM obtains images approximately four times faster on GPUs, compared to parallel executions on traditional CPU-only computing nodes.

6.6 Summary

Light scattering techniques have been widely used for measurement of surface topography and optical critical dimensions. With the fast development of manufacturing techniques, scattering methods have met new applications and new systems have been designed. The development of scattering models and computational technologies, such as machine learning and parallel computing, play an important role to make the scattering technique applicable in more applications in precision and semiconductor manufacturing. The technique continues to develop towards characterisation of more complex surfaces with non-periodic structures, applications that require in-process monitoring of the manufacturing process in a harsh environment and applications where high accuracy and resolution are required.

References

Bawuah P, Pääkkönen P and Peiponen K-E 2017 Gloss measurement in detection of surface quality of pharmaceutical tablets: a case study of screening of genuine and counterfeit antimalaria tablets *J. Eur. Opt. Soc. Rapid. Publ.* **13** 1–7

Beckmann P and Spizzichino A 1987 *The Scattering of Electromagnetic Waves From Rough Surfaces* (Boston, MA: Artech House)

Bennett J and Mattsson L 1999 *Introduction to Surface Roughness and Scattering* (Washington, DC: Optical Society of America)

Cameron A C and Windmeijer F A 1997 An *R*-squared measure of goodness of fit for some common nonlinear regression models *J. Econometrics* **77** 329–42

Coupland J M and Nikolaev N I 2019 A new approach to vector scattering: the 3s boundary source method *Opt. Express* **27** 30380–95

Delacrétaz Y, Seydoux O, Chamot S, Ettemeyer A and Depeursinge C 2012 Monte Carlo simulation of the field back-scattered from rough surfaces *J. Opt. Soc. Am.* A **29** 270–7

Diebold A C 2001 *Handbook of Silicon Semiconductor Metrology* (New York: CRC Press)

Dong Z, Chen X, Wang X, Shi Y, Jiang H and Liu S 2019 Dependence-analysis-based data-refinement in optical scatterometry for fast nanostructure reconstruction *Appl. Sci.* **9** 4091

Franz W 1948 Zur Formulierung des huygensschen Prinzips *Z. Naturforsch.* A **3** 500–6

Fujino T, Tanaka Y, Harada T, Nagata Y, Watanabe T and Kinoshita H 2015 Extreme ultraviolet mask observations using a coherent extreme ultraviolet scatterometry microscope with a high-harmonic-generation source *Jpn J. Appl. Phys.* **54** 06FC01

Gross H, Scholze F, Rathsfeld A and Bär M 2009 Evaluation of measurement uncertainties in EUV scatterometry *Proc. SPIE* **7390** 73900T

Harada T, Nakasuji M, Kimura T, Nagata Y, Watanabe T and Kinoshita H 2011 The coherent EUV scatterometry microscope for actinic mask inspection and metrology *Proc. SPIE* **8081** 80810K

Harada T, Nakasuji M, Nagata Y, Watanabe T and Kinoshita H 2013 Phase imaging of EUV masks using a lensless EUV microscope *Proc. SPIE* **8701** 870119

Henn M-A, Gross H, Scholze F, Wurm M, Elster C and Bär M 2012 A maximum likelihood approach to the inverse problem of scatterometry *Opt. Express* **20** 12771–86

Herffurth T, Schröder S, Trost M, Duparré A and Tünnermann A 2013 Comprehensive nanostructure and defect analysis using a simple 3D light-scatter sensor *Appl. Opt.* **52** 3279–87

Herffurth T, Trost M, Beier M, Steinkopf R, Heidler N, Pertermann T and Schröder S 2019 Assessing surface imperfections of freeforms using a robotic light scattering sensor *Opt. Eng.* **58** 092609

Hergert W and Wriedt T 2012 *The Mie Theory: Basics and Applications* (Berlin: Springer)

Hidayetoglu M, Pearson C, Chew W C, Gürel L and Hwu W-M 2017 Large inverse-scattering solutions with DBIM on GPU-enabled supercomputers *Proc. Int. Appl. Comput. Electromag. Soc. Symp. (ACES) (Florence, Italy, March 2017)* 1–2

Hidayetoglu M, Pearson C, El Hajj I, Gurel L, Chew W C and Hwu W-m 2018 A fast and massively-parallel inverse solver for multiple-scattering tomographic image reconstruction *Proc. IEEE Int. Parallel and Distributed Process. Symp. (IPDPS) (Vancouver, BC, Canada, May 2018)* pp 64–74

Hmamouchi R, Larif M, Chtita S, Adad A, Bouachrine M and Lakhlifi T 2016 Predictive modelling of the LD50 activities of coumarin derivatives using neural statistical approaches: electronic descriptor-based DFT *J. Taibah. Univ. Sci.* **10** 451–61

Hocheng H, Tseng H, Hsieh M and Lin Y 2018 Tool wear monitoring in single-point diamond turning using laser scattering from machined workpiece *J. Manuf. Process.* **31** 405–15

Imlau M, Bruening H, Voit K-M, Tschentscher J and Dieckmann V 2016 Riblet sensor—light scattering on micro structured surface coatings *arXiv*:1601.04694

Kato A, Burger S and Scholze F 2012 Analytical modeling and three-dimensional finite element simulation of line edge roughness in scatterometry *Appl. Opt.* **51** 6457–64

Kato A and Scholze F 2010 The effect of line roughness on the reconstruction of line profiles for EUV masks from EUV scatterometry *Proc. SPIE* **7636** 7636I

Klein R, Laubis C, Müller R, Scholze F and Ulm G 2006 The EUV metrology program of PTB *Microelectron. Eng.* **83** 707–9

Leach R K 2011 *Optical Measurement of Surface Topography* (Berlin: Springer)

Liu M, Cheung C F, Senin N, Wang S, Su R and Leach R K 2020 On-machine surface defect detection using light scattering and deep learning *J. Opt. Soc. Am.* A **37** B53–9

Liu M, Senin N and Leach R 2019 Defect detection for structured surfaces via light scattering and machine learning *Proc. 14th Int. Symp. Measure. Technol. Intell. Ins. (ISMTII) (Niigata, Japan, September 2019)*

Liu S, Du W, Chen X, Jiang H and Zhang C 2015 Mueller matrix imaging ellipsometry for nanostructure metrology *Opt. Express* **23** 17316–29

Lu R 2016 *Light Scattering Technology for Food Property, Quality and Safety Assessment* (Boca Raton, FL: CRC Press)

Madsen J S M, Jensen S A, Nygård J and Hansen P E 2018 Replacing libraries in scatterometry *Opt. Express* **26** 34622–32

Madsen M H and Hansen P-E 2016a Imaging scatterometry for flexible measurements of patterned areas *Opt. Express* **24** 1109–17

Madsen M H and Hansen P-E 2016b Scatterometry—fast and robust measurements of nano-textured surfaces *Surf. Topogr.* **4** 023003

Madsen M H, Hansen P-E, Zalkovskij M, Karamehmedovi M and Garnæs J 2015 Fast characterization of moving samples with nano-textured surfaces *Optica* **2** 301–6

Matsapey N, Faucheu J, Flury M and Delafosse D 2013 Design of a gonio-spectro-photometer for optical characterization of gonio-apparent materials *Meas. Sci. Technol.* **24** 065901

Moharam M and Gaylord T 1981 Rigorous coupled-wave analysis of planar-grating diffraction *J. Opt. Soc. Am.* **71** 811–8

Mouroulis P and Macdonald J 1997 *Geometrical Optics and Optical Design* (New York: Oxford University Press)

Müller C 2013 *Foundations of the Mathematical Theory of Electromagnetic Waves* (Berlin: Springer)

Neubert J, Seifert T, Czarnetzki N and Weigel T 1994 Fully automated angle resolved scatterometer *Proc. SPIE* **2210** 543–52

Niu X, Jakatdar N, Bao J and Spanos C J 2001 Specular spectroscopic scatterometry *IEEE Trans. Semicond. Manuf.* **14** 97–111

Ogilvy J A 1991 *Theory of Wave Scattering from Random Rough Surfaces* (Bristol: Adam Hilger)

Orji N G, Badaroglu M, Barnes B M, Beitia C, Bunday B D, Celano U, Kline R J, Neisser M, Obeng Y and Vladar A 2018 Metrology for the next generation of semiconductor devices *Nat. Electron.* **1** 532–47

Osten W 2016 *Optical Inspection of Microsystems* (Boca Raton, FL: CRC Press)

Paz V F, Peterhänsel S, Frenner K and Osten W 2012 Solving the inverse grating problem by white light interference Fourier scatterometry *Light Sci. Appl.* **1** e36

Prudêncio R B and Ludermir T B 2003 Neural network hybrid learning: genetic algorithms and Levenberg–Marquardt *Proc. 26th Annu. Conf. Gesellschaft Klassifikation e.V. (Mannheim, July 2002)* 464–72

Ramakrishnan D, Singh T, Purwar N, Barde K, Gulati A and Gupta S 2008 Artificial neural network and liquefaction susceptibility assessment: a case study using the 2001 Bhuj earthquake data, Gujarat, India *Computat. Geosci.* **12** 491–501

Scholze F, Beckhoff B, Brandt G, Fliegauf R, Gottwald A, Klein R, Meyer B, Schwarz U, Thornagel R and Tuemmler J 2001 High-accuracy EUV metrology of PTB using synchrotron radiation *Proc. SPIE* **4344** 402–13

Scholze F, Kato A and Laubis C 2011 Characterization of nano-structured surfaces by EUV scatterometry *J. Phys.* **311** 012006

Scholze F, Laubis C, Buchholz C, Fischer A, Plöger S, Scholz F and Ulm G 2006 Polarization dependence of multilayer reflectance in the EUV spectral range *Proc. SPIE* **6151** 615137

Schröder S, Herffurth T, Blaschke H and Duparré A 2011 Angle-resolved scattering: an effective method for characterizing thin-film coatings *Appl. Opt.* **50** C164–71

Schröder S, Unglaub D, Trost M, Cheng X, Zhang J and Duparré A 2014 Spectral angle resolved scattering of thin film coatings *Appl. Opt.* **53** A35–41

SEMI 2009a *Guide for Angle Resolved Optical Scatter Measurements on Specular or Diffuse Surfaces SEMI ME 1392-1109* (Milpitas, CA: Semiconductor Equipment and Materials International)

SEMI 2009b *Test Method for Measuring the Effective Surface Roughness of Optical Components by Total Integrated Scattering SEMI ME 1048-1109* (Milpitas, CA: Semiconductor Equipment and Materials International)

Simonsen I 2010 Optics of surface disordered systems *Eur. Phys. J. Spec. Top.* **181** 1–103

Simonsen I, Maradudin A A and Leskova T A 2010 Scattering of electromagnetic waves from two-dimensional randomly rough penetrable surfaces *Phys. Rev. Lett.* **104** 223904

Stover J C 2012 *Optical Scattering: Measurement and Analysis* 3rd edn (Bellingham, WA: SPIE Optical Engineering Press)

Taflove A and Hagness S C 2005 *Computational Electrodynamics: the Finite-Difference Time-Domain Method* (Boston, MA: Artech House)

Thomas M, Su R, Nikolaev N, Coupland J and Leach R K 2020 Modelling of interference microscopy beyond the linear regime *Opt. Eng.* **59** 034110

Von Finck A, Herffurth T, Duparré A, Schröder S, Lequime M, Zerrad M, Liukaityte S, Amra C, Achour S and Chalony M 2019 International round-robin experiment for angle-resolved light scattering measurement *Appl. Opt.* **58** 6638–54

Zhu H, Lee Y, Shan H and Zhang J 2017 Maximum contributed component regression for the inverse problem in optical scatterometry *Opt. Express* **25** 15956–66

Zhu J, Jiang H, Shi Y, Zhang C, Chen X and Liu S 2015 Fast and accurate solution of inverse problem in optical scatterometry using heuristic search and robust correction *J. Vac. Sci. Technol.* B **33** 031807

Zhu J, Jiang H, Zhang C, Chen X and Liu S 2019 Application of support vector machine for the fast and accurate reconstruction of nanostructures in optical scatterometry *arXiv:*1905.06857

Chapter 7

In-process surface topography measurements

Wahyudin P Syam

In this chapter, general aspects about optical in-process surface topography measurement will be discussed, including clear definitions of terms related to types of measurement. The chapter will include discussions about the effects of environmental noise (such as thermal variation and vibration) on the accuracy of in-process measurement results, challenges to develop fast and accurate optical in-process instruments, and state-of-the-art solutions or developments (such as the use of machine learning algorithms) to solve these challenges and recent advances. A general methodology to develop fast and accurate optical in-process measuring instruments is presented. Finally, a future outlook for the development of in-process measuring instruments concludes this chapter.

7.1 Introduction

High-throughput manufacturing requires measurement systems for produced parts that are accurate and fast enough so that they do not reduce production rates (Takaya 2014, Gao *et al* 2019, Leach 2020). Surface inspection of parts is critical because the surface carries important information about the functionality with reference to the design intent, for example, friction resistance, wear resistance, efficient surface contact, and hydrophobic and hydrophilic properties (Zhang *et al* 2015, Brinksmeier *et al* 2020).

However, the implementation of in-process measurement into production machines is still limited due to the lack of fast and accurate measuring instruments that can operate in harsh environments. To be considered fast, the measurement speed of an in-process instrument should be less than or at least equal to the cycle time of a process into which the instrument will be integrated, so that it does not slow down the overall manufacturing process (Uhlmann *et al* 2016). In discrete manufacturing systems, the process cycle time of the production machines can be less than one minute or even a few seconds. The process cycle time is even faster for production machines in continuous manufacturing systems, such as roll-to-roll

doi:10.1088/978-0-7503-2528-8ch7

processes. Furthermore, to encourage industry uptake, the cost of in-process instruments, including their integration cost, should be lower than the cost of the production machines into which they will be integrated.

This chapter discusses in-process surface topography measurement for parts of millimetre to micrometre dimensions, with millimetre to micrometre feature sizes, and for parts with tight feature tolerances produced by micro-scale manufacturing processes (Tosello 2019) or by other precision manufacturing methods.

The development of in-process surface topography measuring instruments requires challenges to be addressed that are not only related to the measurement speed, but also to other issues, such as environmental noise, which can reduce the quality of measurement results, data fusion and system-level integration. Considering the requirement to achieve fast measurement speed, tactile (contact) instruments are often not suitable. Although accurate, traceable and well-understood, tactile methods can be significantly slower than optical methods, obtain a lower number of points on the surface of a part and have a risk of damaging the surface (Fang *et al* 2013, Gameros *et al* 2015, Uhlmann *et al* 2016). For these reasons, in many situations, optical instruments are more suitable for in-process measurement due to their advantages over tactile instruments, including the ability to obtain high-density data on a part surface within a relatively short measurement time, to gain access to complex geometries and to measure surfaces without the risk of damage (Fang *et al* 2013). However, there are many challenges that hinder the development of in-process optical instruments for measurements performed inside a production machine and/or during a manufacturing process.

This chapter focuses on various aspects related to optical measuring instruments for in-process surface topography measurement, starting with definitions of terms related to where and when a surface topography measurement is carried out. Then environmental issues that can negatively affect the results of in-process measurements are discussed. Development challenges related to the measurement methods, measurement speed, system integration and control, traceability and intelligence to build in-process measuring instruments are also presented, along with recent developments and implementations for in-process surface topography measurement and a general methodology to develop fast and accurate optical measuring instruments. Finally, a future outlook for in-process instrument development is presented.

7.1.1 Definitions

There are many terms that have been developed by researchers and industrialists to refer to their specific instrument development and manufacturing process. To avoid confusion and ambiguity, several terms are given below following definitions developed as part of a recent UK roadmap (Leach 2020):

- *In-process measurement* is any measurement that is performed inside a manufacturing process chain, either outside or inside a production machine (manufacturing station). The measurement results from this type of measurement are used to detect or predict issues in the process and to monitor and control the process. This measurement type includes *in-line, on-machine* and *in situ* measurement.

- *In-line measurement* is a form of in-process measurement and is performed between production machines (manufacturing stations), right before or after a process.
- *On-machine measurement* is a form of in-process measurement and is performed inside a production machine (manufacturing station).
- *In situ measurement* is a form of on-machine measurement and is performed at the location in which a manufacturing process occurs.
- *Off-line measurement* is any measurement that is not in-process and is not synchronised with a manufacturing process so that the measurement cannot be used for process monitoring and control. *Off-line measurement* is commonly performed in a controlled laboratory or in a measurement station outside a manufacturing process/line.

It is clear that from the definitions above, in-process measurement is performed inside or during a manufacturing process chain. The measurement results from in-process measurement and its subsets can be used to monitor and control a process. A pictorial representation of the definitions is shown in figure 7.1.

7.2 Environmental issues

In-process surface topography measurements typically have more uncertainty influence quantities than off-line measurements. Environmental factors, for example, the variation of temperature, pressure, humidity and vibration magnitude, as well as contamination, that are insignificant or can be controlled in the case of off-line measurements, become highly influential for in-process measurement. These environmental factors often cannot be controlled during in-process measurements, in particular during a production process. Additional environmental noise that may have to be considered for in-process measurement can be due to electric and magnetic fields (Weckenmann and Bernstein 2013), for example, on a capacitive displacement sensor (Bohl *et al* 2019). Disturbances created from these fields can reduce the accuracy of the sensor due to coupling at metallic surfaces (Bohl *et al* 2019).

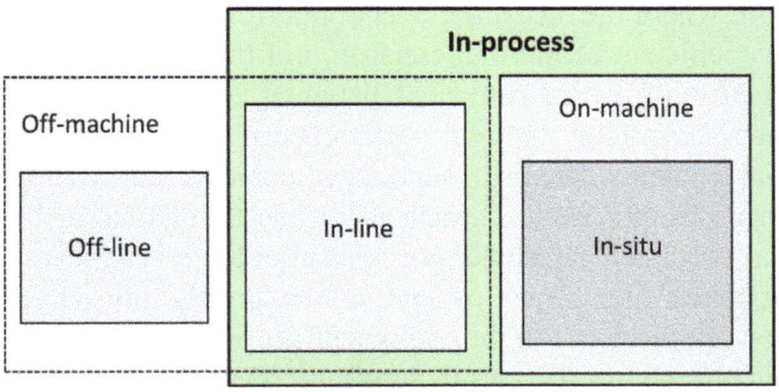

Figure 7.1. Definitions of terms related to in-process, in-line, on-machine, *in situ* and off-machine measurements (adapted from Leach 2020).

The most common environmental effect on in-process measurement is changes in the environmental temperature, causing precision components, such as optical mounts, to expand. Temperature changes also cause the measured part to expand (Schwenke *et al* 2008, Acko *et al* 2015). Due to these environmental noise sources, estimating the 'task-specific' uncertainty for in-process measurement results is more difficult than for off-line measurements because more influence quantities need to be considered (Wilhelm *et al* 2001, Morse 2018).

7.2.1 Temperature, pressure and humidity variation

The variation in temperature, pressure and humidity of an environment where measurements requiring sub-micrometre to nanometre accuracy are carried out may significantly affect the accuracy of the results. Examples of such measurements are interferometry systems or fringe projection, that require phase analysis of signals and a knowledge of the effective wavelength of light. Environmental variations will change the effective refraction index of the medium in which the light travels (Ciddor 1996, Bönsch and Potulski 1998). This change in refractive index will cause a change in the effective wavelength of the light and subsequently affect measured distances. In addition, air turbulence in the room where measurements are carried out will have an effect on the accuracy of the measurement, as air turbulence can also affect the refractive index of a medium and subsequently affect the effective wavelength (Yamauchi and Hibino 2019). Air turbulence is common in a workshop and in a machining chamber.

Although the variations in ambient pressure and humidity do not directly affect the length of a measured part or the mechanical structure of an instrument, they can affect the results of length measurements. For distance measurement with a laser interferometer having a visible laser source, a change in ambient pressure of 1 hPa and relative humidity of 1% will contribute to a measured distance uncertainty of 0.27 and 0.009 μm, respectively, per metre of length (Haitjema 2018). These uncertainties are clearly relevant for distance measurements requiring sub-micrometre to nanometre accuracy.

Related to temperature variations, the dimensions of measured parts and the mechanical structure of the measuring instruments used will expand as the temperature of the room where the measurements are performed increases. This expansion depends on the coefficient of thermal expansion of the materials of the part and the mechanical structure of the instrument (Chetwynd 2018).

On a machine shop floor, thermal variation can come from many sources, for example, the heat generated from surrounding equipment, such as electrical motors on machines, and from the process itself, such as the injection moulding heating process or material removal processes. Machine tools will generate a significant amount of heat from cutting processes. This generated heat will be directly transferred to the machine structure, to the work piece and to the components of an in-process instrument via both conduction and convection processes (Schmitt and Peterek 2015). The heat generated inside a machine tool chamber can be as high as 50 °C which is more than double the standard temperature for dimensional measurement of 20 °C (Berger *et al* 2015).

The effect of temperature variation on the length of a measured part can be reduced by compensation methods, either real-time or model-based off-line compensation (Bohl and Knapp 2019). Often, the compensation will be effective when the temperature changes gradually. Hence, recording ambient and work piece temperatures when possible is useful to understand the effect of the temperature variation and when to apply compensation.

7.2.2 Vibration

Vibration can have a significant effect on measurement results, in particular for measurements at sub-micrometre to nanometre accuracy (Elmadih *et al* 2020). Vibration can arise from many sources with different frequency and magnitude. For example, low frequency vibration can come from ground motion and air conditioning fans and high frequency vibration can come from motor spindle rotation, actuation systems and the bearings of electric motors on motion stages. Inside a manufacturing line with operating production machines, vibration will come from many difference sources and can have relatively high magnitude. In a typical workshop with machine tools, the vibration frequency is in the range 50–200 Hz (Elmadih *et al* 2018), and for in-process optical measurement, this vibration can significantly deteriorate the accuracy of measurement results.

Lower frequency vibration can also degrade measurement results where sub-micrometre to nanometre accuracy is required. For example, in a controlled measurement laboratory, the fan of an air conditioner (AC) generates low frequency vibration. An example vibration spectrum is shown in figure 7.2 (Barker *et al* 2016). The dominant frequency at around 15 Hz is due to AC fan rotation, but there are other sources, for example, a vacuum pump and another measuring instrument in the same room rotating at 1450 RPM (resulting in a peak at 24 Hz).

The low frequency vibration significantly increases the measurement noise of an interferometer by up to 40% (Barker *et al* 2016). It was found that the average measurement noise of the interferometer was 36 nm when the fan was on and 26 nm when off. Figure 7.3 shows the effect of the low frequency vibration noise on a reconstructed surface topography measured by the interferometer, showing the strip-like noise on the measured surface.

When considering the condition inside production machines, vibration noise will be larger than in a controlled laboratory. Hence, this noise will have a more significant effect on the measurement results of optical instruments. Vibration noise can be reduced, if not eliminated, by reducing the magnitude of vibration coming from a source (for example, moving parts, ground vibration and rotating machines), isolating the vibration from the source and isolating an instrument from the vibration source.

Another source of low frequency vibration is the motion of an air-bearing stage when finding its equilibrium position. This vibration can be caused by non-optimal control of the motion stage when adjusting the air supply of the air-bearing to reach a steady position (Wang *et al* 2012). It has been shown that this low frequency

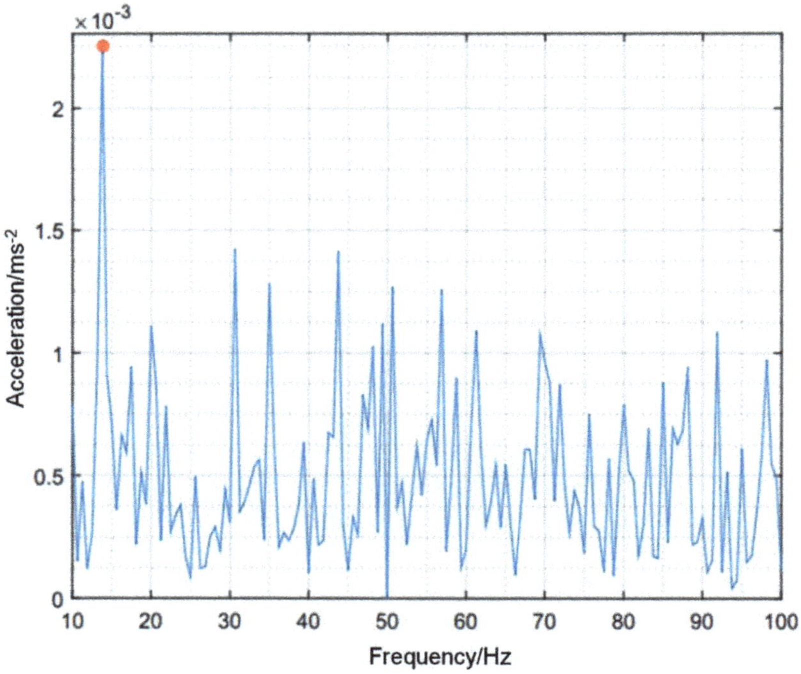

Figure 7.2. Frequency spectrum of the measured vibration signal from a sensor placed on the measurement table of an interferometer (Barker *et al* 2016, copyright ASPE).

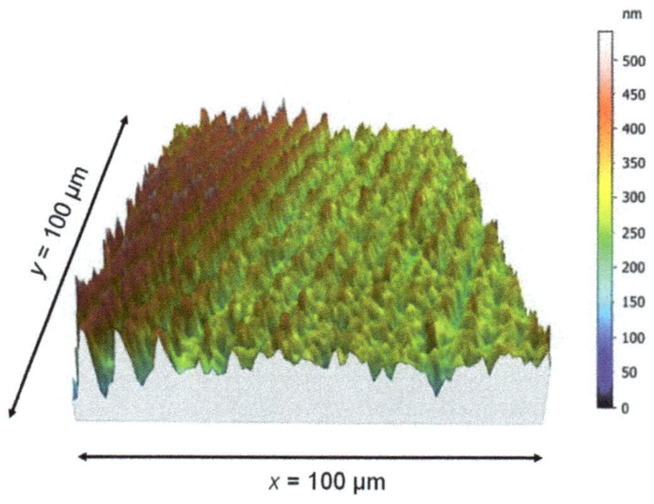

Figure 7.3. The effect of low frequency vibration noise originated from the fan blower of the AC to a reconstructed surface topography measured by an interferometer (Barker *et al* 2016, copyright ASPE).

vibration causes an increase in measurement noise of an on-machine focus variation microscope (FVM) from 0.37 μm measured in a controlled lab to 0.73 μm measured in the machining chamber of an precision machine tool while its *xy*-motion stage was active (Santoso *et al* 2020).

There are many solutions to vibration isolation, which are discussed elsewhere (Elmadih *et al* 2018). One recent solution is to optimise the shape of the mechanical structure of an instrument using lattice structures, so that the structure can provide vibration isolation and damping. Structures can be manufactured using additive manufacturing (AM), for example, laser powder bed fusion (LPBF). AM processes enable the manufacture of parts with high complexity, such as the lattice structures, that can have less mass than bulk material structures and properties to damp and isolate vibration (Syam *et al* 2018a, Elmadih *et al* 2019a). For vibration reduction, one solution can be reducing the speed of the motor of a motion stage (de Groot and Di Sciacca 2018). In many implementations, solutions to reduce vibration are specific and depend on the type of instrument and the condition of the measurement environment. For example, Muhamedsalih *et al* (2010) used a piezo-transducer attached to the reference mirror of an on-machine wavelength scanning interferometer to compensate for fluctuations in the optical length caused by vibration from the surrounding environment.

7.2.3 Contamination

Contamination, such as dust, dirt and films, is common in manufacturing processes; it can be from the removed material from a manufactured part or from the environment of a production line. Dust, dirt and films can deteriorate the measurement accuracy of optical instruments as they can adhere to the surfaces of the optical components and affect light transmission/reflection in the optical train (Weckenmann and Bernstein 2013). This effect, for example, can reduce the quality of images from optical instruments and change the properties of optical components. Contamination can be a significant issue as in-process measurements are often performed in uncontrolled environments compared to a clean measurement laboratory. In addition, if contaminants attach to the surface of measured parts, they can be measured and cause invalid features on the reconstructed surface topography, in turn affecting the calculation of surface texture parameters.

Hence, the enclosure design for in-process instruments should be able to mimimise solid contaminant ingress, but also reduce liquid generated in the form of coolant oil or other fluids. It is important that the enclosure design should comply with the 'IP' (waterproofing) rating for at least dust and liquid protection levels (IEC 1989)—conforming to the IP68 standard is a good recommendation for the enclosure specification. Figure 7.4 shows an example of an in-process measurement condition where dust and dirt from the machined part could be significant and may affect the measured surface. In figure 7.4, the instrument is close to the machined part where debris can be floating around the chamber and attach to the part or the optical components.

7.3 Development challenges

Developing fast and accurate optical in-process surface topography measuring instruments requires solutions to challenges that are not necessarily significant for the development of off-line instruments. These challenges are because more factors

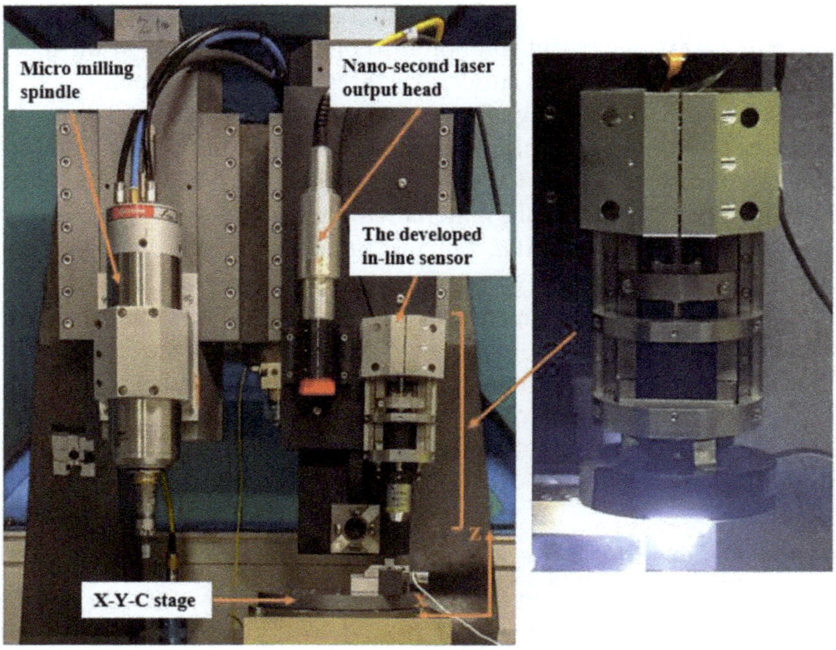

Figure 7.4. An example of an environment of in-process measurement with dust and dirt from removed materials (Santoso *et al* 2020, copyright ASPE).

affect in-process measurements and contribute to their measurement uncertainty. The challenges can be categorised into five types: methods, speed, system integration and control, traceability, and intelligence. Here, the definition of intelligence is the ability of an instrument to learn from prior measurements and improve its future measurement performance, in terms of data coverage, accuracy and speed.

Some factors are relevant not only for optical instruments, but also for tactile instruments. For example, the challenge to perform multi-scale measurements, that is, measurements at hundreds of millimetre scales for form and at sub-micrometre scales for texture, and to measure in noisy environments, in particular for measurements requiring sub-micrometre to nanometre accuracy, are relevant for both optical and tactile measurements.

It is worth noting that, during the development of an optical in-process measuring instrument, not all challenges need to be overcome. Rather, only those challenges that are significant for the specific situation in which the instrument is developed need to be addressed. Table 7.1 presents a summary of issues related to each type of challenge and a longer description is given in the following sections.

7.3.1 Measurement method

The measurement method challenge is related to the limitations of optical measuring instruments, some of which are discussed here. The maximum surface slope angle that can be measured and optical lateral resolution limit are fundamental physical limitations of optical instruments equipped with objective lenses (i.e. acting as

Table 7.1. Challenges for implementing optical in-process measuring instruments.

	Challenge	Current state-of-the-art
Method	Long measurement time for large surface area.	Multiple measurements of small surface areas followed by data stitching procedures (Leach 2011).
	Wide spatial bandwidth (multi-scale) measurements for form and surface texture.	Either measurement of a large area with low resolution or a small area with high resolution (Leach *et al* 2013, 2015b).
	Measurement of high slope angle features.	• High slope angle feature measurement limitation by the numerical aperture (NA) of an objective lens for microscope-based instruments (Leach and Haitjema 2010). • Current high slope angle measurement beyond NA with certain configurations, advanced measurement models, advanced algorithms, such as vertical focus probing (Zangl *et al* 2018) and using *a priori* information (Leach *et al* 2018c).
	3D surface reconstruction degradation due to specular reflection from reflective surfaces.	3D reconstruction problems due to the saturation of the pixels of an imaging sensor caused by high reflection from a surface (Leach 2011).
	Reduced measurement accuracy under a noisy environment, for example, vibration and temperature variation.	Significant negative effects of environmental noise, such as ground vibration, for measurements with micrometre and higher-level accuracies (Barker *et al* 2016). For an example, the differences between an encoder reading and the actual position during a measurement caused by small levels of vibration.
Speed	Measuring faster than a process cycle time.	Larger measuring time (commonly > one minute) than process (commonly seconds) cycle time for areal surface measurements due to the requirement to access a surface and a large number of image acquisitions and computations (Leach 2011, 2016).

(Continued)

Table 7.1. (*Continued*)

	Challenge	Current state-of-the-art
	Short processing time for handling and analysing high-density data from optical measuring instruments.	• The time required to process the data is more than the time to acquire it. • A large number of data points, from hundreds of thousands (Zhang 2010) to millions or more points (Senin *et al* 2017), can be obtained from optical measuring instruments in relatively short periods of time (Leach 2011, 2016).
System integration and control	The need of applications of modular and environmentally robust design to integrate and adapt into various types of manufacturing machines.	• Different types of machines have their own requirements for specifically built in-process measuring instruments due to, for example, space constraints and level of vibration on the machines (Mendikute *et al* 2018, Li *et al* 2018a, Gardner *et al* 2018). • Some proposals to passively isolate vibration by using lattice structures (Syam *et al* 2018a, Elmadih *et al* 2019b, 2019c). The implementation of a lattice structure metal frame for vibration isolation has been reported (Leach *et al* 2020).
	Algorithms for utilisation and combination of various types of in-process measurement data, from different sensors with different resolutions, for the efficient control of a manufacturing system.	• The availability of many sensors with different resolutions and accuracy levels to a gather large number and variety of data. • The use of data fusion to combine all the data with different densities (Weckenmann *et al* 2009). • Several reported works on fusion of data with different densities, for example, data fusion from two optical instruments (Leach *et al* 2018a) and data fusion from a tactile and an optical instrument (Colosimo *et al* 2015, Hocken and Pereira 2012).

	The need for integration of an in-process instrument with real-time system-level control (for example, on-line statistical process control, run-to-run control and predictive maintenance).	• The availability of several methods for system-level control (Chien *et al* 2013). • Batch mode, instead of continuous mode, run-to-run process controls application due to the lack of in-line instruments (Liu *et al* 2018, Lu *et al* 2018).
	The need for integration of measurement data into an enterprise production planning and scheduling.	Non-unified data infrastructure between measurement data from in-line and/or in-process instruments and data in the resource planning management system, used for production planning and scheduling, of companies and enterprises (Higashide *et al* 2010).
	High-speed data transfer to avoid data bottlenecks.	Bottlenecked data transfer due to large amounts of data, originated from hardware or software, transferred with limited speeds (Wollschlaeger *et al* 2017).
Traceability	The need of the infrastructure for calibration and verification of performance of in-line measuring instruments to assure measurement traceability and the instruments work within their specification.	• Lack of performance verification procedures and material measures for the determination of dimensional and surface measurement errors for optical instruments (Balcon *et al* 2012, Leach *et al* 2015a). • Currently, the availability of performance verification infrastructures is only for some off-line optical instruments (Claverley and Leach 2015, Moroni *et al* 2014, 2017). • The quantification of metrological characteristic to calibrate optical measuring instrument (ISO 25178–600 (ISO 2019)). In some situations, the calibration infrastructures of off-line optical instruments are applied to in-line optical instruments for surface measurements (Leach *et al* 2015a, 2015c, Giusca *et al* 2012a, 2012b, Giusca and Leach 2013, Alburayt *et al* 2018, Leach 2018b, Biro and Kinnel 2020).

(Continued)

Table 7.1. (*Continued*)

	Challenge	Current state-of-the-art
	Methods and procedure for uncertainty contributor quantification for the estimation of the measurement uncertainty associated with in-line measurement results to establish measurement traceability.	Lack of current methods and procedure of measurement uncertainty estimation, commonly applied only for off-line tactile and optical measuring instruments (Tosello *et al* 2009, Moroni *et al* 2018, Sims-Waterhouse *et al* 2020).
Intelligence	The utilisation and application of machine learning methods for the performance improvement of in-line optical measurements.	• A recent application of deep learning in fringe projection measurement to rapidly tracking the projector orientation has been reported (Stavroulakis *et al* 2019). • Some applications of deep learning for object classifications from 3D point clouds have also been reported (Monti *et al* 2017, Wang *et al* 2018).
	The requirement of effective training of a model from very large data sets, obtained from in-line optical measurements, within a relatively short period of time.	• The use of a parallel computation methods leveraging graphical processing units (Krizhevsky *et al* 2017). • The use drop-out (Srivastava *et al* 2014) and penalisation (Pereyra *et al* 2017, Krizhevsky *et al* 2016) regularisation methods to avoid overfitting of large training data sets and to increase accuracy of deep learning methods.
	The need for uncertainty estimation with machine learning methods as, very often, machine learning methods are used as 'black-box' methods.	• The use of a Bayesian framework with deep learning to provide uncertainty estimation for deep learning (Gal *et al* 2017, Peharz *et al* 2018, Karaletsos *et al* 2018). • The use of prediction uncertainty to obtain confidence on results and to infer the reasonability of the results.

microscopes). These fundamental limits depend on the numerical aperture (NA) of the objective lenses. Furthermore, optical instruments using an objective lens have limited fields of view or measuring areas (typically on millimetre and sub-millimetre scales) and relatively short working distances between the objective lens and a measured part surface (Leach 2011). The 3D surface reconstruction of many optical instruments can be negatively affected by saturation on their imaging sensor caused by light intensities that exceed the pixel threshold value, typically from highly reflecting surfaces (Leach 2011). As discussed in the above, optical measurements at sub-micrometre to nanometre accuracy are significantly affected by environmental noise, such as vibration, temperature, humidity and pressure variations, and these variations significantly contribute to the resulting measurement uncertainty (Haitjema 2018).

7.3.2 Measurement speed

The measurement speed challenge (this can also be considered a measurement time challenge and will often be quantified in units of time) does not only refer to the speed to collect raw data, but also to the speed of motion to access the desired features on a surface and to the speed of storing, handling and processing the high-density data commonly obtained from optical instruments. The total measurement speed of any measuring instruments, including tactile instruments, is determined by how fast they can capture and reconstruct raw data, for example a stack of images of a surface, and how fast a measurement result can be derived from the raw data. In particular for optical instruments that can produce high-density data, reducing the total measurement speed, including measuring a surface, reconstructing its 3D surface model and calculating specific parameters from the reconstructed surface data, becomes more challenging than for tactile instruments.

7.3.3 System integration and control

The system integration and control challenge includes the integration of an in-process surface topography measuring instrument into the process chamber of a production machine and the integration of its measurement results into a system-level data management system. With system-level integration of the measurement results, a production system can be monitored and controlled by an integrated data management system at an organisational level. These abilities to monitor and control the production system become a significant economic advantage of in-process measuring instruments. The economic advantages can be obtained through the improvement of efficiency and competitiveness of manufacturing organisations (Schleich *et al* 2018). A common method, used to monitor and control a production system is statistical process control (SPC), which can be used to detect whether a process or a part deviates from its predefined condition or nominal design. However, currently, SPC methods are primarily used to process data off-line or per production batch (but see section 7.4).

Controlling a production system on-line is challenging as it needs fast and accurate in-process measuring instruments. These instruments should perform measurements in uncontrolled environments with noise, in particular during processes inside production machines, and within a short time period (less than or equal to the production bandwidth). In addition to the speed and accuracy requirements of in-process instruments, their integration ability in a production machine highly depends on their physical design. Subsequently, modular design concepts need to be adapted so that the instruments can adapt to various production machines regardless of, for example, space constraints.

7.3.4 Traceability

Traceability is a fundamental requirement for measurement and includes calibration and performance verification of in-process optical measuring instruments and the uncertainty estimation associated with their measurement results (see definitions in chapter 1). There has been a significant amount of research to develop material measures and procedures to verify the performance of off-line optical measuring instruments (Leach *et al* 2015a), but less for in-process measuring instruments (but see Kumar *et al* 2016, Jones and O'Connor 2018). ISO 25178 part 600 (ISO 2019) provides guidelines on how to calibrate an optical measuring instrument by determining a defined set of metrological characteristics, which are applicable to all types of optical measuring instruments. The main issue related to the estimation of measurement uncertainty for optical instruments is not on what methods should be used to estimate the uncertainty, but the identification and quantification of relevant influence factors for an in-process optical instrument (Gao *et al* 2019). For a more detailed discussion on the traceability and calibration infrastructure for optical measuring instruments, see Leach *et al* (2015a).

7.3.5 Intelligence

Intelligence for measuring instruments means that the instruments can learn from prior information and/or prior measurements to improve their measurement performance, by using machine learning (ML) methods. The performance improvement can be, for example, more data coverage, higher resolution and accuracy of measurement results, and shorter measurement times. Currently, ML methods, although not yet common, have been implemented for measurements, for example, the use of ML methods to understand surface orientations (Stavroulakis *et al* 2019), to infer surface information from missing data using *a priori* information (Senin and Leach 2018), and to automatically segment reconstructed 3D point clouds obtained from optical instruments (Monti *et al* 2017).

Recently, deep neural network or deep learning methods have found application in many fields, from natural and social sciences, engineering to business. Deep learning uses neural networks with many neurons and layers, and multiple hierarchies of abstractions (LeCun *et al* 2015). In engineering, deep learning methods are intensively used for machine vision applications to automatically segment and identify objects in images (LeCun *et al* 2015, Chen *et al* 2018). The fast proliferation and widespread use of deep learning methods is due to the availability of abundant data used for training, affordable computers with high memory capacity and computing power, and advanced learning algorithms to find optimum ML parameters. These methods perform significantly better than conventional neural networks with only a few layers and far less neurons (Krizhevsky *et al* 2012) and other classical ML methods, for example, support vector machines (Bishop 2013).

One of the main drawbacks of deep learning methods is the lack of capability to estimate the uncertainty of predictions (Gal and Ghahramani 2016), as these methods are often used as black-box systems. This drawback causes a fundamental issue for their implementation in measurement applications. In measurement, the estimation of measurement uncertainty is required not only to establish traceability, but also to aid decision taking for part conformance in quality control (Haitjema 2018).

To address the problem to estimate prediction uncertainty in ML in general, and deep learning in particular, Bayesian probabilistic approaches may be able to address the uncertainty prediction problems (Ghahramani 2015). However, Bayesian approaches require higher memory capacity than neural network methods (Bishop 2013). Subsequently, the combination of Bayesian approaches and deep learning may add the capability of deep learning methods to be able to estimate uncertainty of their prediction (Gal and Ghahramani 2016).

7.4 Recent advances and developments

7.4.1 Recent advances in instrument development

Judging by the recent research literature, measuring instruments based on interferometry are the most widely adopted systems for in-process surface topography measurement. Despite their high sensitivity to environmental variations, interferometric methods have advantages of height resolution that allows sub-micrometre to nanometre accuracy, the ability to be configured in many different ways using different illumination sources, for example laser or white-light sources, and different methods for signal analysis. These various configurations and signal analysis methods can be selected to comply with a specific measurement requirement and functionality.

Autschbach *et al* (2010) have developed a phase-shifting interferometry (PSI) system based on a Twymann-Green configuration and integrated it into an ultra-precision turning machine to measure the complex surface of a manufactured part (figure 7.5(a)). Their system is compact and can be integrated into the machining chamber with simple fixtures. Figure 7.5(b) shows the integrated PSI on the turning machine. A dispersed reference interferometry (DRI) system integrated into an ultra-precision turning machine has been developed (Li *et al* 2018a). DRI uses the machine axes to reach a desired location for measurement (figure 7.5(c)). To improve the accuracy of the DRI measurement, Li *et al* also calibrated and quantified the kinematic parameters of the machine axes (Li *et al* 2018b).

A wavelength scanning interferometer (WSI) using two Linnik objectives has been developed for in-process surface topography measurements without performing axial scanning along its optical axis (Jiang *et al* 2010) (see chapter 5 for a review

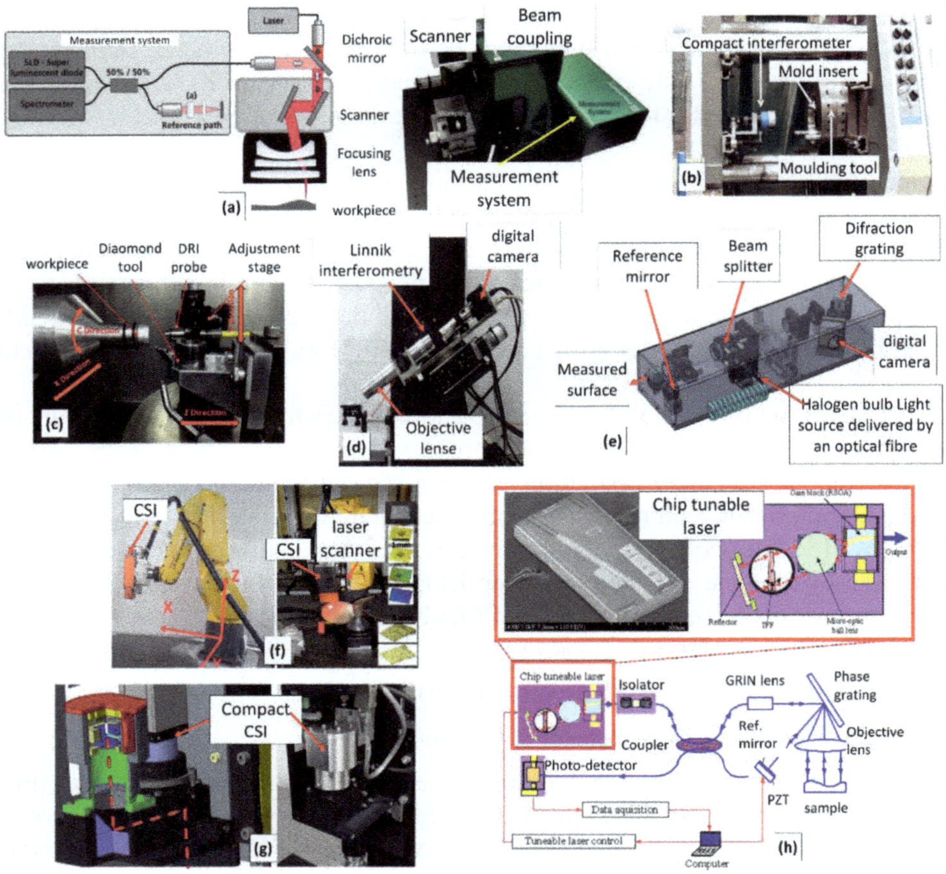

Figure 7.5. (a) Twymann-Green PSI mounted on to a laser surface texturing machine (Schmitt *et al* 2013, with permission from Elsevier), (b) PSI system based on a Twymann-Green configuration integrated into an ultra-precision turning machine (Autschbach *et al* 2010), (c) DRI system integrated into an ultra-precision turning machine (Li *et al* 2018a, copyright Euspen), (d) WSI using two Linnik objectives for in-process surface topography measurements without performing axial scanning along its optical axis (Jiang 2011, with permission from Elsevier), (e) white-light channelled spectrum interferometer without needing to be at a focus position for measurement (Jiang *et al* 2015, with permission from Elsevier), (f) CSI integrated into an industrial robot for in-process measurement (Biro and Kinnel 2020, copyright IOP Publishing, reproduced with permission, all rights reserved), (g) compact CSI sensor integrated into the machining chamber of a laser surface texturing machine (Bermudez *et al* 2019, copyright Euspen) and (h) compact laser interferometer in a single chip that consists of a tunable laser, directional coupler, optical isolator and photodetector (Yang *et al* 2010, copyright IOP Publishing, reproduced with permission, all rights reserved).

of non-axial scanning optical instruments). To improve the performance of the WSI, an active vibration compensation system, using a piezo-electric actuator and control system, is applied to the reference mirror of the WSI (Martin *et al* 2008, Jiang *et al* 2006). To increase the measurement speed of the WSI, real-time analysis of the signal correlogram is run on a graphics processing unit (Jiang 2011). This high-speed WSI has been installed on a large drum diamond turning machine. Figure 7.5(d) shows the high-speed WSI mounted on a motion stage. Another application of the

WSI was for in-process measurement in roll-to-roll production of thin films (Muhamedsalih *et al* 2015). One of the drawbacks of the WSI is the need for finding a focus position before measurement. To overcome this drawback, a white-light channelled spectrum interferometer (WLCSI) has been developed (Jiang *et al* 2015, Gao *et al* 2015). This WLCSI does not require to be at a focus position to perform measurements, so that surface measurements can be performed faster than with WSI (figure 7.5(e)), although there may be issues with how defocus affects the spatial frequency response on the instrument. Yang *et al* (2010) have developed a compact laser interferometer in a single chip that consists of a tunable laser, directional coupler, optical isolator and photodetector. With this single-chip tunable laser, a compact interferometry system can be built. Furthermore, this interferometer is designed to have a self-stabilised system to reduce vibration noise (figure 7.5(h)).

Among all interferometry systems, coherence scanning interferometry (CSI) is the most used method for in-process measurements (see chapter 2). Schmitt *et al* (2012, 2013) have developed a CSI based on a Twymann-Green configuration and mounted it on to a laser surface texturing machine. This CSI is used to measure the surface form of an ultra-precision mould insert. Their integration can be achieved by using the same optical path for both the interferometry system and the laser system used for surface texturing (figure 7.5(a)). A CSI with an anti-vibration system has been developed by Chen *et al* (2012). This CSI uses an additional external laser interferometer to measure the small movement, due to vibration, of the objective and compensate the movement. Biro and Kinnel (2020) integrated a CSI into an industrial robot by mounting it on to the robot's end effector so that the CSI can be used for in-process measurement. Figure 7.5(f) shows the commercial CSI mounted on the robot's end effector. Biro and Kinnel evaluate the measurement noise of the in-process CSI and, based on this evaluation, design strategies and demonstrate the use of the CSI for in-process measurement of hole geometries and surface texture in simulated manufacturing settings. A compact CSI sensor, with narrow bandwidth white-light source and up to 1 nm height resolution, has been developed and integrated into the machining chamber of a laser surface texturing machine (Bermudez *et al* 2019). The CSI can measure the surface form of a laser textured surface within 2 s and the result is used to compare the topography of the textured surface to the nominal design. The difference between the nominal and textured surface is used to correct the laser tool path (figure 7.5(g)).

Hovell *et al* (2020) developed a lens-less fibre-coupled optical coherence tomography (OCT) sensor that can be used for in-process absolute height measurements (figure 7.6(a)). To reduce the sensitivity of the instrument to vibration, thermal and humidity variations, they use a fully fibre enclosed system, that is compact and can be mounted beside the nozzle of an electrochemical jet machining system. The fibre system delivers both sample and reference signals of a broadband light source to a part surface. Their instrument can achieve an accuracy of 76 nm for up to an 8 μm step height measurement in non-ideal (uncontrolled) environments. Gardner *et al* (2018) developed in-process OCT, with 11.7 μm height resolution and up to 3.36 mm subsurface depth measurement, for a polymer LPBF process. This OCT system is

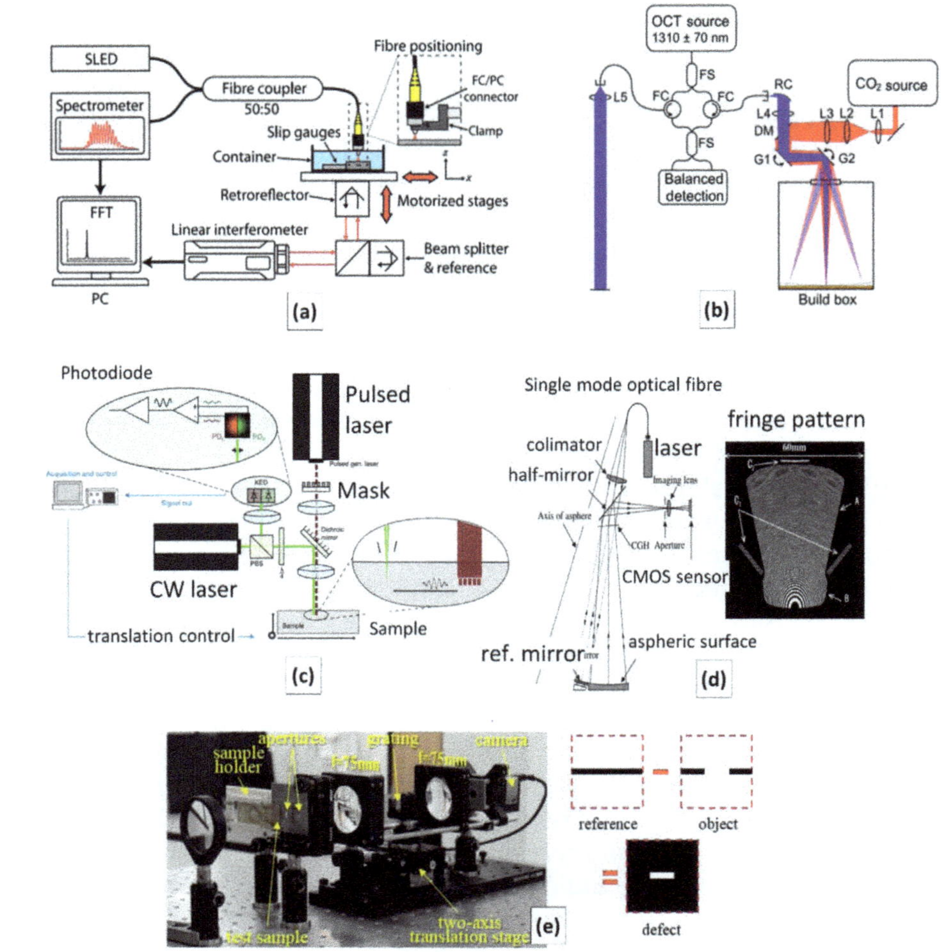

Figure 7.6. (a) Lens-less fibre-coupled OCT sensor used for an in-process absolute depth/height measurements (Hovell *et al* 2020, copyright SPIE), (b) in-process OCT for LPBF process (Gardner *et al* 2018, copyright SPIE), (c) spatially resolved acoustic spectroscopy method to image surface topography for in-process measurement of LPBF process (Patel *et al* 2018, copyright MDPI), (d) digital holography method for in-process measurement of the surface of a large 1.2 m diameter off-axis spherical mirror (Kino and Kurita 2012, copyright OSA) and (e) AODE used for in-process measurement of printed-electronics with roll-to-roll processes.

used to monitor change and detect defects on a surface and subsurface of melted powder (figure 7.6(b)).

Other developments for in-process surface topography measurement have been reported. The use of spatially resolved acoustic spectroscopy (SRAS) method to image surface topography has been developed for in-process measurement of an LPBF process (Patel *et al* 2018). Their system measures the just-built layer and determines whether the quality of the build layer is within specification (figure 7.6(c)). Kino and Kurita (2012) implemented a digital holography method for in-process measurement of the vertical and lateral surface of a large off-axis spherical mirror with a diameter of 1.2 m. This holography system was installed on a large surface grinding machine which processed the mirror (figure 7.6(d)). Feng *et al* (2018) developed an all-optical difference engine (AODE) for in-process measurement of printed-electronics with roll-to-roll processes. The prototype AODE can measure an area of 4 mm × 4 mm with several micrometre lateral resolution. The high-speed inspection of the AODE was achieved by minimising data processing and by comparing a measured surface to a 'golden standard' surface (figure 7.6(e)).

In addition to interferometry systems, several commercial and in-house developed FVMs have been reported (see more examples in chapter 3). Nasrollahi *et al* (2017) integrated a commercial FVM sensor into the machining chamber of a micro-scale laser machine tool. This integration was used to measure drilled micro-holes using a novel two-sided laser processing method (figure 7.7(a)). Danzl *et al* (2017) mounted a commercial FVM sensor on to the end effector of a collaborative robot to measure micro-geometries on a large part, while the part was still inside a machine tool and on the machining fixture. The mounting of the FVM sensor on the robot is shown in figure 7.7(b). A development of a prototype compact FVM sensor that can be integrated into various machine tools for on-machine measurement has been reported (Darukumalli *et al* 2019, Santoso *et al* 2020). This compact FVM sensor has dimensions of 80 mm diameter and 200 mm length, a height resolution of 20 nm and requires less than 20 s for a single image-field surface topography measurement (figure 7.7(e)).

A compact imaging confocal microscopy (ICM) sensor (see chapter 4) that does not require in-plane scanning over a surface has been developed for in-line surface topography measurement (Martinez *et al* 2020). The ICM uses structured illumination to project a high-resolution and deterministic spatial frequency pattern to recover the bright field of the obtained sectioned images of a surface. With this illumination system, the measurement time of the ICM can be significantly reduced by eliminating the in-plane scanning process. However, the ICM still requires vertical scanning along its optical axis. With the reduced measurement time and compact design, the ICM can potentially be used for in-process measurement (figure 7.7(c)). Quinsat and Tournier (2012) integrated a compact chromatic confocal sensor for in-process surface texture measurement. The sensor is integrated into a milling machine to inspect profile and areal surfaces after the machining process (figure 7.7(d)).

Other methods of fringe projection, photogrammetry, deflectometry and laser triangulation, that do not have the NA limitation found in systems with microscope objectives, have been developed. Fringe projection profilometry (FPP) systems can capture a large area of a surface and produce high-spatial resolution data (Chen *et al* 2020). For these reasons FPP systems have been integrated into an LPBF machine and roll-to-roll manufacturing systems. Börchers and Gromov (2008) have implemented an FPP system for in-process surface flatness measurement on flat-rolled products. They demonstrate the effectiveness of the FPP system for the flatness measurement of steel plates, sheets and strips in the steel industry (figure 7.8(a)). Dickins *et al* (2020) have developed a multi-view FPP solution for in-process measurement of LPBF (Dickins *et al* 2020). The goal of the multi-view system is to obtain images with more coverage and minimise the effects of specular reflection

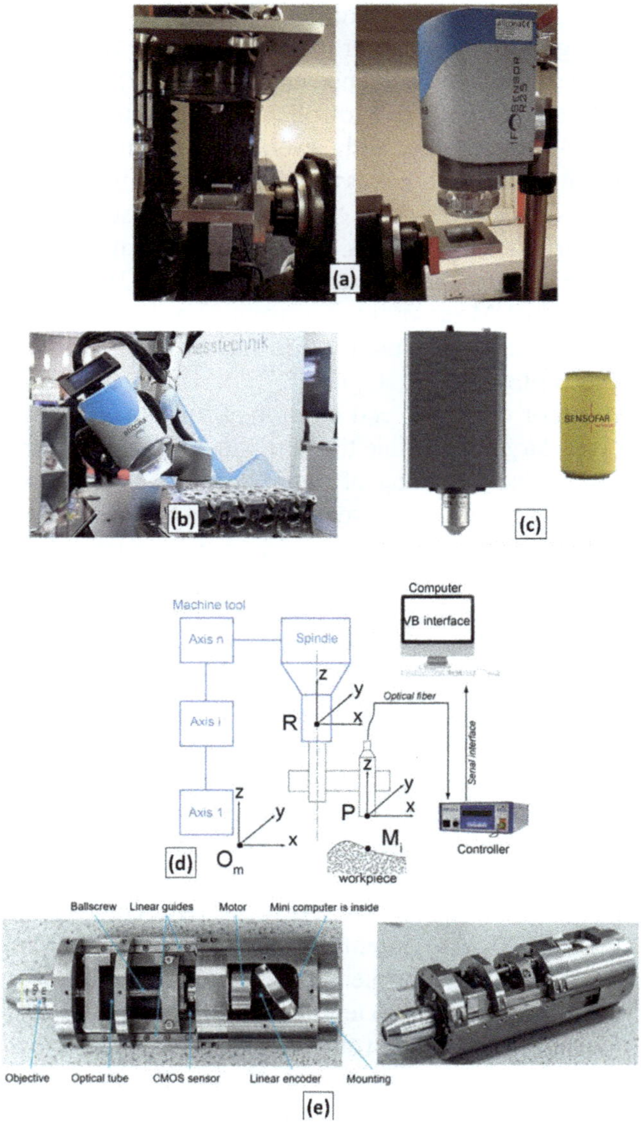

Figure 7.7. (a) Integration of a commercial FVM sensor into the machining chamber of a micro-scale laser machine tool (Nasrollahi *et al* 2017, with permission from AMSE), (b) mounting of a commercial FVM sensor on to the end effector of a collaborative robot to measure micro-geometries on a large part (Danzl *et al* 2017, with permission from EDP Sciences), (c) compact ICM sensor that does not require in-plane scanning over a surface (Martinez *et al* 2020, copyright SPIE), (d) compact chromatic confocal sensor integrated into a milling machine for in-process surface texture measurement (Faber *et al* 2012, copyright Euspen) and (e) new development of a compact FVM sensor that can be integrated into various machines (Santoso *et al* 2020, copyright Springer Nature).

(figure 7.8(b)). Zhang *et al* (2016) have developed a custom-designed in-process FPP system, shown in figure 7.8(c), to measure the surface topography of powder beds and characterise process signatures. A similar FPP set-up to that of Zhang *et al* has

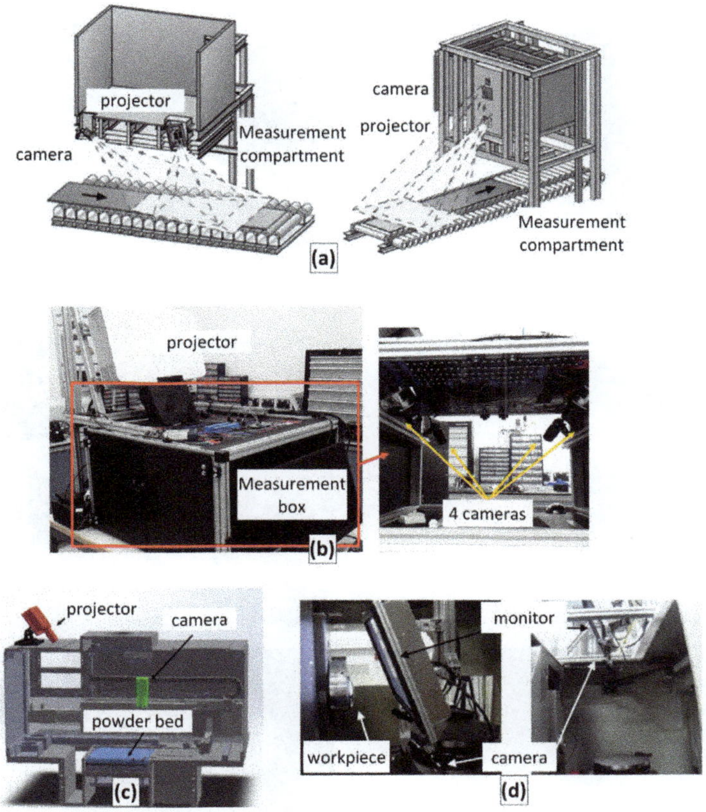

Figure 7.8. (a) FPP system for in-process surface flatness measurement on flat-rolled products (Börchers and Gromov 2008, copyright Springer Nature), (b) solution for in-process measurement of LPBF process with multi-view FPP (Dickins *et al* 2020, copyright OSA), (c) custom-designed in-process FPP system to measure the surface topography of powder beds and their process signature (Zhang *et al* 2016, with permission from Elsevier) and (d) custom-designed deflectometry system integrated into the machining chamber of an ultra-precision polishing machine tool for measuring highly reflective surfaces (Faber *et al* 2012, copyright Euspen).

also been developed by Liu *et al* (2018b) and has been integrated into an electron beam powder bed fusion machine. A custom-designed deflectometry system has been integrated into the machining chamber of an ultra-precision polishing machine tool for measuring highly reflective or mirror like surfaces (Röttinger *et al* 2011, Faber *et al* 2012). Deflectometry has a similar surface reconstruction method to FPP and can achieve up to a few nanometres of accuracy (Häusler *et al* 2013) but captures the pattern of structured fringes reflected from a surface from a flat panel. By interrogating the reflected pattern, the fringes on a specular surface can be captured to reconstruct the 3D surface model (figure 7.8(d)). The performance of deflectometry has been improved by combining several measurements to achieve accuracies up to 400 nm for surface height measurement of highly reflective surfaces (Olesch and Häusler 2014).

Photogrammetry systems (Sims-Waterhouse *et al* 2020) have been used for in-process measurement due to their relatively simple set-up. Common set-ups for these

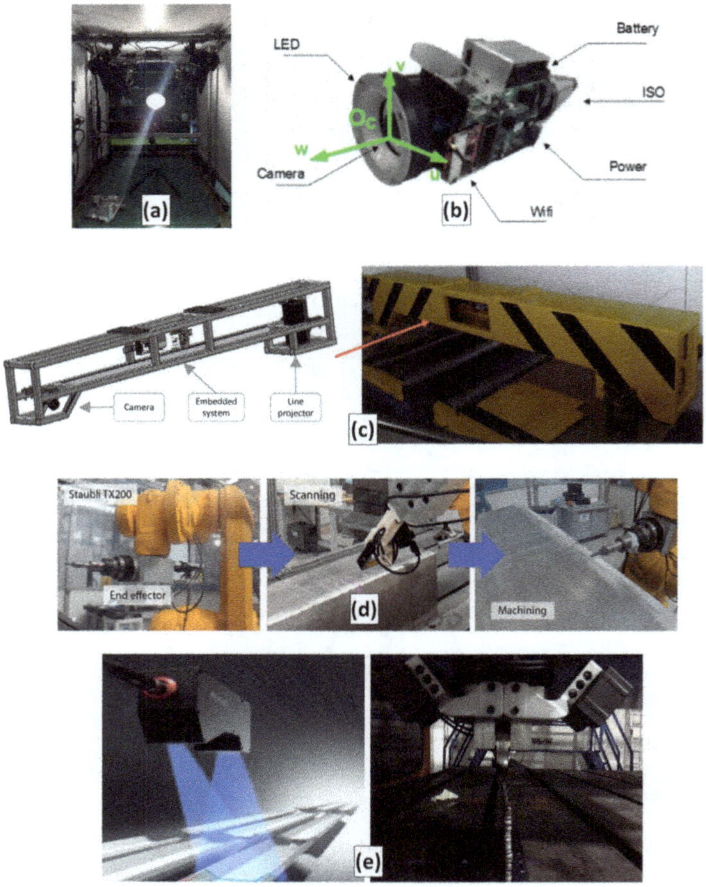

Figure 7.9. (a) Photogrammetry system with two cameras for in-process surface form measurement of parts on a conveyor (Bergström *et al* 2016), (b) single camera photogrammetry system mounted on the spindle of a machine tool (Mendikute *et al* 2018, with permission from Elsevier), (c) laser scanning system for in-process measurement of the flatness of rolled steels (Molleda *et al* 2013, with permission from Elsevier), (d) laser scanning system mounted on the end effector of a robot for in-process surface form measurements for robotic finishing process (Gurdal *et al* 2019, with permission from Elsevier) and (e) an in-process laser scanning system for monitoring the build process of a wire-arc-based additive machine (Xu *et al* 2018, copyright Taylor and Francis).

systems use multiple cameras or a single camera with multiple image captures at different positions. Bergström *et al* (2016) demonstrated the use of a photogrammetry system with two cameras (figure 7.9(a)) for in-process surface topography measurement of parts on a conveyor to control their geometrical quality. A single camera photogrammetry system mounted on the spindle of a machine tool has been developed (Mendikute *et al* 2018). This single camera system has a compact design and can be moved anywhere within the machine tool working volume to measure the surface topography of a part (figure 7.9(b)).

Molleda *et al* (2013) developed a laser triangulation system for in-process measurement of the flatness of rolled steels. They use a typical laser scanning configuration with enhanced hardware and software flatness measurement in less than 1 s (figure 7.9(c)). A similar laser triangulation system for flatness measurement of rolled steels was developed (Bilstein 2014). A laser triangulation system was mounted on the end effector of a robot to perform in-process surface topography measurements for robotic finishing processes (Gurdal *et al* 2019). The laser system, with 12 μm lateral resolution and 100 mm axial and lateral scan areas, was used to measure the topography of a stir-welded part and the measurement result is fed into the robot controller to perform finishing processes and to remove burrs from the welded part (figure 7.9(d)). Another development of an in-process instrument using laser triangulation has been proposed (Xu *et al* 2018). They mounted the laser scanning system on the cladding head of a wire-arc-based AM machine to monitor the build process (figure 7.9(e)).

Machine vision is commonly used for in-process measurement, although it can only capture the 2D image of a surface. An in-process surface topography measurement combining machine vision and laser triangulation has been proposed (Czajka *et al* 2018). The instrument was integrated into a turning machine to measure the

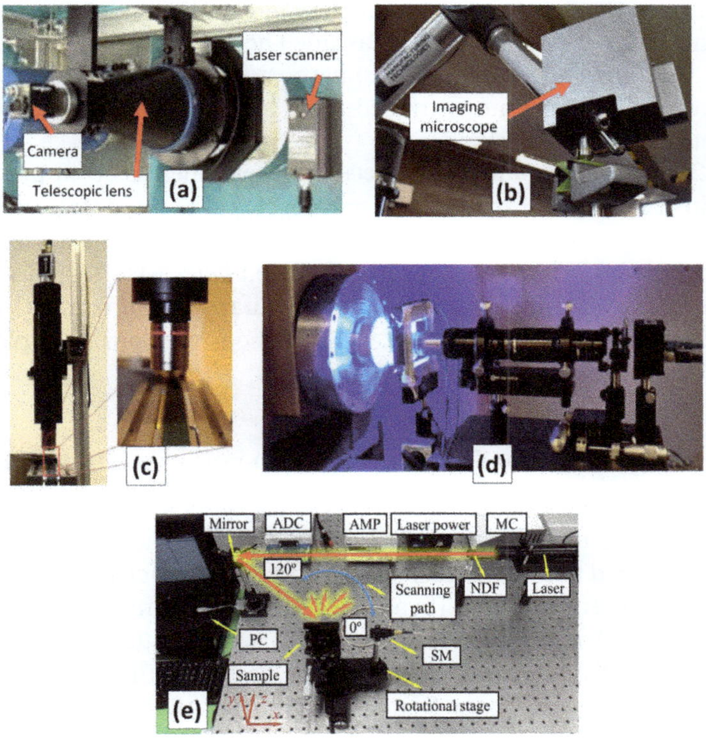

Figure 7.10. (a) In-process surface topography measurement combining machine vision and laser scanning system (Czajka *et al* 2018, copyright SPIE), (b) compact imaging microscope enhanced with a machine learning algorithm for 2D surface topography measurement (Syam *et al* 2019, with permission from Elsevier), (c) (Weimer *et al* 2014, with permission from Elsevier) and (d) imaging microscope based on planoptic (light-field) camera for in-process surface topography measurement on a micro-scale cold forming system (Li *et al* 2014, with permission from Elsevier) and (e) light scattering instrument to determine classes of deterministic structured surfaces and estimate their geometric properties (Liu *et al* 2020, copyright SPIE).

topography of a machined cylinder, for example, diameter, run-out and profile texture of the surface (figure 7.10(a)). A compact imaging microscope enhanced with an ML algorithm has been proposed (Syam *et al* 2019). The imaging microscope was used for in-line measurement of the surface texture of a post-process additive surface and can be potentially used for in-process measurement due to its compact design (figure 7.10(b)).

Weimer *et al* have developed an imaging microscope based on a planoptic (light-field) camera (figure 7.10(c)). This microscope can obtain both profile and areal surface topography of metallic micro-parts and has been integrated into a micro-scale cold forming system (Weimer *et al* 2014). Li *et al* (2014, 2015) developed a stereoscopy 3D measurement system using a planoptic configuration for in-process measurement of structured surfaces (figure 7.10(d)). The stereoscopy system is integrated into an ultra-precision machine tool.

A light scattering instrument to determine classes of deterministic structured surfaces and estimate their geometric properties has been developed (see chapter 6 for a review of scattering methods). This light scattering instrument can be used for in-process surface topography measurement (Liu *et al* 2020). The scattering instrument uses an ML algorithm based on a cascaded neural network to process scattered signals and determine the class and geometric properties of a structured surface (figure 7.10(e)). This algorithm is trained using simulated scatter signals generated from a boundary element method.

Table 7.2 shows a summary of recent developments of in-process optical measuring instruments. Despite the many developments reported, 67% of them are still in working prototype phases and need further development. Interferometry methods account for 38% of all methods discussed here. Most applications for the in-process measurement are for micro-scale manufacturing processes, for example, laser texturing and grinding/polishing. FPP methods are mostly implemented for powder bed layer measurements in AM processes.

7.4.2 Methodology to develop fast and accurate in-process measuring instruments

A general methodology to develop in-process measurement systems, including in-line and on-machine instruments, based on ‘information-rich metrology’ (IRM), a framework developed at the University of Nottingham (Leach *et al* 2018c, Senin and Leach 2018), has been proposed (Syam *et al* 2019). This methodology consists of three phases: phase 1 is for knowledge and data (*a priori*) gathering; phase 2 is for instrument and control software development, based on the knowledge and data gathered in phase 1; and phase 3 is the utilisation of the developed in-process measuring instrument (in phase 2) for in-line control (figure 7.11).

The IRM framework is a fundamental element of the methodology to develop in-process instruments (Leach *et al* 2018c, Senin and Leach 2018) and refers to the exploitation of any available relevant information to improve the performance of a measurement process (Leach *et al* 2018c, Senin and Leach 2018). The relevant information can be, for example, information related to the expected geometry or surface texture of a measured part obtained from its CAD model, the ‘fingerprint’ of

Table 7.2. The summary of recent developments of in-process optical measuring instruments.

Source	Method	Application	FoV	Resolution	Speed	Accuracy and performance
Autschbach *et al* (2010)	PSI	Micro-scale turning machining	NA	NA	NA	Peak-to-valley error = 10–20 nm (repeatability 3.2 nm), measurement noise = 2 nm (repeatability 0.5 nm)
Li *et al* (2018a, 2018b)	DRI	Micro-scale turning machining	Point measurement (point-to-point scanning)	Axial resolution = 0.6 nm (800 μm vertical range)	NA	Step height error = 90 nm
Jiang *et al* (2010), Jiang (2011), Muhamedsalih *et al* (2015)	WSI	Roll-to-roll process	(0.5 × 0.7) mm with 5× objective lens	Axial resolution = 0.05 nm	Measurement cycle = 9 s	Step height error = 1 nm
Jiang *et al* (2015), Gao *et al* (2015)	WLCSI	Roll-to-roll process	(0.5 × 0.7) mm with 5× objective lens	Axial resolution = 0.05 nm	Measurement cycle = 1–2 s	Step height error = 20 nm
Yang *et al* (2010)	Chip tuneable laser	Roll-to-roll process	NA	Lateral resolution = 4.65 μm	Data acquisition time = 60 s	Length measurement error = 3.21 nm, measurement repeatability = 1.96 nm
Schmitt *et al* (2012, 2013)	CSI	Micro-scale laser surface texturing	(30 × 30) mm with in-plane scanning using a f-theta lens	Axial resolution = 4.58 μm	Measurement cycle = 60 s	Measurement error = 218 nm for 1.25 mm length
Biro and Kinnel (2020)	CSI	Industrial robotic	(5.86 × 5.62) mm for 5× objective lens, (1.47 × 1.41) mm for 10× objective lens	Vertical resolution = 100 nm, lateral resolution = 5 μm (10× objective lens) or 20 μm (5× objective lens)	Measurement cycle 60 s or more	Measurement noise = 0.2 μm, residual flatness <0.2 μm
Bermudez *et al* (2019)	CSI	Micro-scale laser surface texturing	<(0.7 × 0.7) mm with 20× Mirau objective lens	NA	Measurement cycle = 2 s	Measurement noise < 100 nm

(*Continued*)

Table 7.2. (*Continued*)

Source	Method	Application	FoV	Resolution	Speed	Accuracy and performance
Hovell *et al* (2020)	OCT	Laser sintering AM	Point measurement (point-to-point scanning)	Axial resolution = 100 nm	NA	Step height measurement error = 56 nm (in air) and 76 nm (in water)
Gardner *et al* (2018)	OCT	Laser melting AM	Point measurement (3.93 μm spot radius)	Axial resolution = 11.7 μm	Data acquisition time for B-scan = 3 s, A-scan frequency = 100 kHz	NA
Patel *et al* (2018)	SRAS	Laser melting AM	Point measurement with 8 μm diameter spot	NA	Scan speed = 1000–1500 points/s	NA
Kino and Kurita (2012)	Holography	Ultra-precision grinding	(1730 × 1388) mm	Lateral resolution = 2.8 mm	NA	Peak-to-valley error < 150 nm
Feng *et al* (2018)	AODE	Roll-to-roll process for printed-circuit roll-roll process	(4 × 4) mm	Lateral resolution = several micrometres	<1 s	Defect-to-noise ratio = up to 74.8
Nasrollahi *et al* (2017), Danzl *et al* (2017)	FVM	Micro-scale laser drilling	(2 × 2) mm	Vertical resolution = 100 nm, lateral resolution = 1 μm	Measurement cycle < 60 s	NA
Santoso *et al* (2020), Darukumalli *et al* (2020)	FVM	Micro-scale milling and laser machining	(1.7 × 1.7) mm with 10× objective lens	Lateral resolution = 0.8 μm, vertical resolution = up to 20 nm	Measurement cycle < 20 s	Step height measurement error = 20 nm, measurement noise = 0.4 μm
Martínez *et al* (2020)	CM	NA	NA	NA	Measurement cycle < 60 s	Measurement noise = 15.4 nm (10×), 7.5 nm (20×), 2.2 nm (50×), 3.8 nm (150×)
Quinsat and Tournier (2012)	CCS	Milling machine	(4 × 1) mm	Axial resolution = 5 nm	4 s	Dimensional error = 5 μm

Börchers and Gromov (2008)	FP	Flat-rolled process	NA	NA	NA	Flatness measurement error = 100 μm
Dickins *et al* (2020)	FP	Laser melting AM	(250 × 250) mm	Lateral resolution < 50 μm	NA	NA
Zhang *et al* (2016)	FP	Laser melting AM	(28 × 15) mm	Lateral resolution = 6.8 μm	NA	Single point repeatability = 0.47 μm
Liu *et al* (2018b)	FPP	Electron beam melting	(180 × 180) mm	NA	Data acquisition time = 5 s	Length measurement error = 6.9 μm
Röttinger *et al* (2011), Faber *et al* (2012), Olesch E and Häusler G (2014)	Deflectometry	Ultra-precision polishing	(120 × 120) mm	NA	NA	Height measurement error = 470 nm
Bergström *et al* (2016)	Photogrammetry	Conveyor belt process	(60 × 40) cm	Lateral resolution = 84 μm	Measurement cycle = 3 s to 3 min	Depth accuracy = 12–60 μm, lateral accuracy = 5–25 μm
Mendikute *et al* (2018)	Photogrammetry	Milling machine	NA	Lateral resolution = 2.2 μm	NA	Length measurement error = 1 μm
Molleda *et al* (2013)	Laser scanning	Flat-rolled process	NA	Lateral resolution = 70 μm	Measurement cycle < 1 s	Length measurement error = 0.8 μm
Bilstein (2014)	Laser scanning	Flat-rolled process	NA	NA	Measurement speed = 1500 m/min	5% *Ra* deviation from a stylus measurement
Gurdal *et al* (2019)	Laser scanning	Industrial robotic machining	(100 × 100) mm	Lateral resolution = 12 μm	Scanning frequency = 300 Hz	Diameter measurement accuracy = 0.125 mm
Xu *et al* (2018)	Laser scanning	Wire-arc AM	(120 × 120) mm	NA	NA	Length measurement error = 5 μm
Czajka *et al* (2018)	Machine vision + laser scanning	Turning process	(70.9 × 70.9) mm	Lateral resolution = 50 μm	Sampling frequency 16 KHz	Measurement repeatability = 10 μm
	Machine vision			Lateral resolution = ±1 λm		Pixel noise = 2.2 pixel unit

(*Continued*)

Table 7.2. (*Continued*)

Source	Method	Application	FoV	Resolution	Speed	Accuracy and performance
Syam *et al* (2018b, 2019)		AM surface post-processing	(1.2 × 1) mm with 10× objective lens		Measurement cycle < 4 s	
Weimer *et al* (2014)	Plenoptic	Micro-scale cold forming	NA	NA	Image acquisition frequency = 24–140 Hz	Length measurement error = 10 μm
Li *et al* (2014, 2015)	Plenoptic	Ultra-precision turning	(70 × 70) μm	Lateral resolution = 6.75 μm	Measurement cycle = 5 s	Height measurement error < 1 μm
Liu *et al* (2020)	Scattering	Structured surface	Point measurement	NA	Measurement cycle = < 5 s	NA

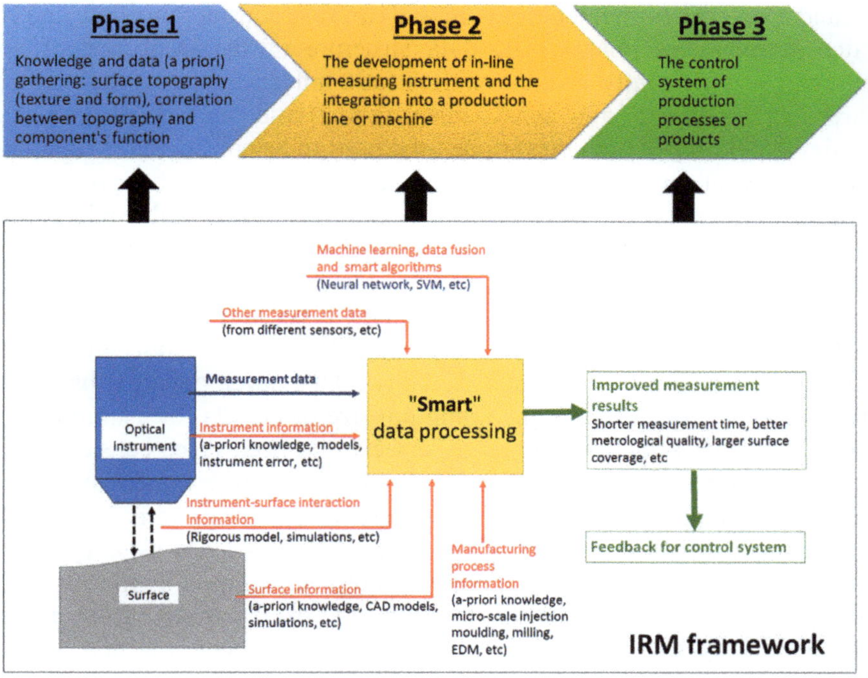

Figure 7.11. The proposed methodology to develop in-process measuring instruments based on the IRM framework: phase 1 is knowledge and data (*a priori*) gathering, phase 2 is instrument and control software development, and phase 3 the utilisation of the developed in-process measuring instruments for in-line control (Syam *et al* 2019, with permission from Elsevier).

a manufacturing process that stays on the manufactured part, the physics of the instrument–surface interaction, the model of a measurement method, the characteristics of the optical instrument used for measuring, the history from previous measurements and any other *a priori* knowledge. All or some of this relevant information can be combined and exploited by the use of a 'smart data processing' to improve the performance of future measurement processes and results (figure 7.11). Various methods and algorithms, for example, ML and multi-scale data fusion are needed to perform the smart data processing to process all the information.

The three phases of the methodology are briefly explained as follows (figure 7.11). Phase 1, 'Knowledge and data (*a priori*) gathering', is to define the measurand for in-process measurement and gather the information that is useful for the IRM framework to develop a fast and accurate in-process measuring instrument. To gather the information, studies of the measurement method to be used and the most important defects on a part to be measured as fast as possible are required. These studies are often carried out using a high-specification (commonly off-line) instruments to gather data about the surface topographies and defects of the part. In this phase, measurement can be time consuming. However, the time invested in this phase is worth the effort as the main goal is to gather accurate information about the measurand and the measurement process, and can save considerable time and cost in the later development phases. The data structure of the measurement results, procedures for uncertainty estimation (relevant influence factors that affect the measurement uncertainty and measurement model) and the type of data analysis (and training data for ML models) to be used need to be considered. From phase 1, the most relevant properties (small changes in values of the properties will significantly affect the part functionality) of a part to be measured will be identified.

Phase 2, 'Instrument and software development', is to develop the 'simplest' yet effective, efficient, fast and accurate in-process measuring instrument utilising the IRM framework for the required measurement tasks and the integration into a production machine where the measurement will be carried out. In this phase, 'specificity' is more important than 'versatility'. This development is based on the results of phase 1, for example, the minimum size of defects (measurand) that are relevant to the part's quality. Important aspects that need to be considered in this phase are speed requirements, instrument cost, accuracy requirements, types of sensor, size constraints, modular design and the choice of programming language for the instrument's software. The most important aspect is speed as, very often, the developed in-process instrument will only be accepted and used by industry if it can measure faster or at least equal to the cycle time of a manufacturing process of interest. To justify the economic benefit of the instrument, the cost should be significantly lower than the cost of the manufacturing process. The instrument should achieve the required accuracy and should not go beyond requirements to keep the cost at a minimum. The instrument design should have overall dimensions that can fit within a given space constraint and be robust to environmental noise. The programming language used for the instrument's software for control and data analysis should be fast and highly compatible with various types of hardware. The C/C++ language may be the most suitable language for this application due to its fast running time and high compatibility to many control systems (Stroustrup 2010). Currently, although slower than C/C++, the Python programming language is becoming popular for instrument software due to its compatibility with different instrument peripherals and support for the Internet-of-Things (IoT) protocol (Von Rossum and Drake 2011) for system-to-machine and machine-to-machine communications, which is essential for industry 4.0 (Lasi *et al* 2015). Finally, the enclosure of the instrument should consider the environment in which the instrument will be integrated.

Phase 3, 'Development of in-line control system', is to provide a control system for product/process monitoring that can be as simple as a 'go/no-go' system or an advanced control system leveraging process models, ML models and feedback data from the developed in-process instrument. The control system prevents defective parts going through subsequent processes or to repair the defective parts before going through subsequent processes (Gao *et al* 2019, Liu *et al* 2018). Advanced control systems use SPC to capture assignable events on a controlled process and give an 'alarm' in order to make a corrective action. Classical SPC is commonly used for off-line control. This off-line control often causes parts with defects, that have been produced, to pass to the next stages when a corrective action is undertaken.

Emerging SPC methods for in-line control are based on run-to-run control applied to a small batch production, while the production is still running, so that a corrective action can be carried out faster than when using classical SPC (Liu *et al* 2018). ML methods can be used to develop intelligent SPC methods that can adapt to correlated data from different probability distributions (Jia *et al* 2018). The control system should consider the uncertainty of the measured data to determine the conformance of a part (Morse *et al* 2018, Haitjema 2018) and to make a better decision, selecting an effective corrective action (Alexander *et al* 2018).

7.5 Summary and future outlook

In-process surface topography measurement has become increasingly important for quality control of high-throughput and high-value manufacturing processes without slowing down the processes and creating 'bottleneck' in the manufacturing flow. Despite the importance of in-process measurement, fast and accurate in-process measuring instruments are not widely available and their capabilities are limited to, for example, relatively slower measurement speed than the processes and lower accuracy than off-line instruments due to environment effects.

There are many challenges to be solved to produce fast and accurate, as well as flexible, in-process measuring systems. The difficulty of these challenges increases for the development of 3D and relatively large areas (several millimetres or more) in-process surface topography measuring instruments. Hence, a great deal of collaborative research and development work from interdisciplinary fields needs to be carried out to address these challenges.

In addition to the collaborative research and development work, emerging knowledge and technologies, for example, computing hardware, novel ML models and optimisation algorithms, new progress in optical physics modelling, novel innovative manufacturing processes and innovative designs, should to be exploited to address the challenges. The availability of more affordable high-performance computing power and novel advanced ML algorithms can be used to improve the performance of current or new measuring instruments, in particular their measurement speed. New understanding and modelling methods and light–surface interactions should be used to develop new measurement methods that are fast and highly accurate. Environmental noise that degrades the accuracy of the measurement results from in-process instruments can be isolated by adapting innovative designs leveraged by additive processes.

References

Alburayt A, Syam W P and Leach R K 2018 Lateral scale calibration for focus variation microscopy *Meas. Sci. Technol.* **29** 065012

Acko B, Klobucar R and Acko M 2015 Traceability of in-process measurement of workpiece geometry *Proc. Eng.* **100** 376–83

Alexander A, Kumar M and Walker H 2018 A decision theory perspective on complexity in performance measurement and management *Int. J. Oper. Prod. Manag.* **38** 2214–44

Autschbach L, Ewert D, Gabbia A, Wang H, Blümel T and Boeck W 2010 Integration of interferometry into ultra-precision lathe for tool alignment and surface analysis *Proc. EUSPEN (Delft, The Netherlands, June 2010)*

Balcon M, Carmignato S and Savio E 2012 Performance verification of a confocal microscope for 3D metrology tasks *Quality Acc. Succ.* **13** 63–6

Barker A, Syam W P and Leach R K 2016 Measurement noise of a coherence scanning interferometer in an industrial environment *Proc. ASPE* 594–9

Berger D, Brabandt D and Lanza G 2015 Conception of a mobile climate simulation chamber for the investigation of the influences of harsh shop floor conditions on in-process measurement systems in machine tools *Measurement* **74** 233–7

Bergström P, Fergusson M, Folkesson P, Runnemalm A, Ottosson M, Andersson A and Sjödahl M 2016 Automatic in-line inspection of shape based on photogrammetry *Proc. Int. Swedish Prod. Symp. (SPS16)* 7 1–9

Bermudez C, Martinez P, Cadevall C and Artigas R 2019 High speed, roll to roll coherence scanning interferometry in a laser texturing process *Proc. EUSPEN (Bilbao, Spain, June 2019)*

Bilstein W 2014 Industrial online surface roughness measurement of strip *Millenium Steel* **16** 150–5

Biro I and Kinnel P 2020 Performance evaluation of a robot-mounted interferometer for an industrial environment *Sensors* **20** 257

Bishop C M 2013 *Pattern Recognition and Machine Learning* (Singapore: Springer)

Bohl S and Knapp W 2019 Observer-based compensation of thermal disturbances for linear displacement sensor *Ann. CIRP* **68** 543–6

Bohl S, Weikart S and Wegener K 2019 Enhancing signal quality of capacitive displacement measurements in machine tool environments *J. Manuf. Mater. Process* **3** 76

Bönsch G and Potulski E 1998 Measurement of the refractive index of air and comparison with modified Edlen's formulae *Metrologia* **35** 133–9

Börchers J and Gromov A 2008 Topometric measurement of the flatness of rolled products—the system TopPlan *Metallurgist* **52** 247–52

Brinksmeier E, Karpuschewski B, Yan J, Riemer O and Schönemann L 2020 Manufacturing of multiscale structured surfaces *Ann. CIRP* **69** 717–39

Chen L C, Papandreou G, Kokkinos I, Murphy K and Yuille A L 2018 DeepLab: semantic image segmentation with deep convolutional nets, atrous convolution, and fully connected CRFs *IEEE Trans. Pattr. Analys. Mach. Intell* **40** 834–48

Chen R, Xu J and Zhang S 2020 Digital fringe projection profilometry *Advances in Optical Form and Coordinate Metrology* ed R K Leach (Bristol: IOP Publishing)

Chen L C, Yeh S L, Tapilouw A M and Lee K F 2012 *In-situ* scanning white light interferometry employing dual-sensing configuration and active fringe-locking strategy *Int. J. Nanomanuf.* **8** 40–52

Chetwynd D G 2018 Materials selection in precision mechanics *Basics of Precision Engineering* ed R K Leach and S T Smith (Boca Raton, FL: CRC Press)

Chien C-F, Hsu C-Y and Chen P-N 2013 Semiconductor fault detection and classification for yield enhancement and manufacturing intelligence *Flex. Serv. Manuf. L* **25** 367–88

Ciddor P E 1996 Refractive index of air: new equations for the visible and near infrared *Appl. Opt.* **35** 1566–73

Claverley J D and Leach R K 2015 A review of the existing performance verification infrastructure for micro-CMMs *Precis. Eng.* **39** 1–15

Colosimo B, Pacella M and Senin N 2015 Multisensor data fusion via Gaussian process models for dimensional and geometric verification *Precis. Eng.* **40** 199–213

Czajka P, Garbacz P, Mężyk J, Giesko T and Mazurkiewicz A 2018 The optomechatronic system for automatic quality inspection of machined workpieces *J. Mach. Construc. Maint.* **4** 77–86

Danzl R, Lankmair T, Calvez A and Helmi F 2017 Robot solutions for automated 3D surface measurement in production *18th Int. Congr. Metrology* 15002

Darukumalli S, Santoso T, Syam W P, Helmli F and Leach R 2019 On-machine optical surface topography measurement sensor based on focus variation *Proc. EUSPEN (Bilbao, Spain, June 2019)*

de Groot P and DiSciacca J 2018 Surface-height measurement noise in interference microscopy *Proc. SPIE* **10749** 10749Q

Dickins A, Widjanarko T, Sims-Waterhouse D, Thompson A, Lawes S and Leach R K 2020 Multi-view fringe projection system for surface topography measurement during metal powder bed fusion *J. Opt. Soc. Am.* A **57** B93–B105

Elmadih W, Chronopoulos D, Syam W P, Maskery I, Meng H and Leach R K 2019b Three-dimensional resonating metamaterials for low-frequency vibration attenuation *Sci. Rep.* **9** 1–8

Elmadih W, Nefzi M and Buice E S 2020 Environmental isolation *Basics of Precision Engineering* ed R K Leach and S T Smith (Boca Raton, FL: CRC Press)

Elmadih W, Syam W P, Maskery I, Chronopoulos D and Leach R K 2019a Mechanical vibration bandgaps in surface-based lattices *Add. Manuf.* **25** 421–9

Elmadih W, Syam W P, Maskery I, Chronopoulos D and Leach R K 2019c Multidimensional phononic bandgaps in three-dimensional lattices for additive manufacturing *Materials* **12** 1878

Elmadih W, Syam W P, Maskery I and Leach R 2018 Designing low frequency band gaps in additively manufactured parts using internal resonators *Proc. ASPE, Charlotte, NC*

Faber C H, Kurz M, Röttinger C H, Olesch E, Domingos D, Löwenstein A, Häusler G and Uhlmann E 2012 Two approaches to use phase measuring deflectometry in ultra-precision machine tools *Proc. EUSPEN (Stockholm, Sweden, June 2012)*

Fang F Z, Zhang X D, Weckenmann A, Zhang G X and Evans C 2013 Manufacturing and measurement of freeform optics *Ann. CIRP* **62** 823–46

Feng X, Su R, Happonen T, Liu J and Leach R K 2018 Fast and cost-effective in-process defect inspection for printed electronics based on coherent optical processing *Opt. Express* **26** 13927–37

Gal Y and Ghahramani Z 2016 Dropout as a Bayesian approximation: representing model uncertainty in deep learning *33rd Proc. (ICML, New York)*

Gal Y, Islam R and Ghahramani Z 2017 Deep Bayesian active learning with image data *arXiv*:1703.02910

Gameros A, De Chiffre L, Siller H R, Hiller J and Genta G 2015 A reverse engineering methodology for nickel alloy turbine blades with internal features *CIRP J. Manuf. Sci. Technol.* **9** 116–24

Gao W, Haitjema H, Fang F Z, Leach R K, Cheung C F, Savio E and Linares J M 2019 On-machine and in-process surface metrology for precision manufacturing *Ann. CIRP* **68** 843–66

Gao F, Muhamedsalih H, Dawei T, Elrawemi M, Blunt L, Xiang J, Edge S, Bird D and Hollis P 2015 *In-situ* defect detection systems for R2R flexible PV barrier films *Proc. SPIE* **9622** 96220E

Gardner M R, Lewis A, Park J, McElroy A B, Estrada A D, Fish S, Beaman J J and Milner T E 2018 *In situ* process monitoring in selective laser sintering using optical coherence tomography *Opt. Eng.* **57** 041407

Ghahramani Z 2015 Probabilistic machine learning and artificial intelligence *Nature* **521** 452–9

Giusca C L and Leach R K 2013 Calibration of the scales of areal surface topography measuring instruments: part 3. Resolution *Meas. Sci. Technol.* **24** 105010

Giusca C L, Leach R K, Helery F, Gutauskas T and Nimishakavi L 2012a Calibration of the scales of areal surface topography measuring instruments: Part 1. Measurement noise and residual flatness *Meas. Sci. Technol.* **23** 035008

Giusca C L, Leach R K and Helary F 2012b Calibration of the scales of areal surface topography measuring instruments: part 2. Amplification, linearity and squareness *Meas. Sci. Technol.* **23** 065005

Gurdal O, Rae B, Zonuzi A and Ozturk E 2019 Vision-assisted robotic finishing of friction stir-welded corner joints *Proc. Manuf.* 40 70–6

Haitjema H 2018 Measurement uncertainty ed R K Leach and S T Smith *Basics of Precision Engineering* (Boca Raton, FL: CRC Press)

Häusler G, Faber C, Olesch E and Ettl S 2013 Deflectometry vs interferometry *Proc. SPIE* **8788** 87881C

Higashide M, Nishina K, Hironobu K and Ishii N 2010 *Statistical Process Control for Semiconductor Manufacturing Processes* (Berlin: Springer)

Hocken R J and Pereira P H 2012 *Coordinate Measuring Machines and Systems* (Boca Raton, FL: CRC Press)

Hovell T, Matharu R S, Petzing J, JusthamL and Kinnel P 2020 Lensless fiber-deployed low-coherence interferometer for *in situ* measurements in nonideal environments *Opt. Eng.* **59** 014113

IEC 1989 Degrees of protection provided by enclosures (IP code) *IEC System of Conformity Assessment Schemes for Electrotechnical Equipment and Components—IEC 60529* (Geneva: International Electrotechnical Commission)

ISO 25178 part 600 2019 *Geometrical Product Specifications (GPS)—Surface Texture: Areal—Part 600: Metrological characteristics for areal-topography measuring methods* (Geneva: International Organization for Standardization)

Jia X, Di Y, Feng J, Yang Q, Dai H and Lee J 2018 Adaptive virtual metrology for semiconductor chemical mechanical planarization process using GMDH-type polynomial neural networks *J. Proc. Control* **62** 44–54

Jiang X 2011 *In situ* real-time measurement for micro-structured surfaces *Ann. CIRP* **60** 563–6

Jiang X, Dawei T and Gao F 2015 *In-situ* surface inspection using white light channelled spectrum interferometer *Proc. 15th Met. Prop. (March 2015)*

Jiang X, Wang K, Gao F and Muhamedsalih H 2010 Fast surface measurement using wavelength scanning interferometry with compensation of environmental noise *Appl. Opt.* **49** 2903–9

Jiang X, Wang K and Martin H 2006 Near common-path optical fiber interferometer for potentially fast on-line microscale-nanoscale surface measurement *Opt. Lett.* **31** 3603–5

Jones C W and O'Connor D 2018 A hybrid 2D/3D inspection concept with smart routing optimisation for high throughput, high dynamic range and traceable critical dimension metrology *Meas. Sci. Technol.* **29** 074004

Karaletsos T, Dayan P and Ghahramani Z 2018 Probabilistic meta-representations of neural networks *arXiv*:1810.00555

Kino M and Kurita M 2012 Interferometric testing for off-axis aspherical mirrors with computer-generated holograms *Appl. Opt.* **19** 4291–7

Krizhevsky A, Sutskever I and Hinton G E 2012 ImageNet classification with deep convolutional neural networks *NIPS* 1097–105

Krizhevsky A, Sutskever I and Hinton G E 2016 System and method for addressing overfitting in a neural network *US Patent* US2016/0335540A1

Krizhevsky A, Sutskever I and Hinton G E 2017 System and method for parallelizing convolutional neural networks *US Patent* US9563840B2

Kumar P, Martin H and Jiang X 2016 Towards the development of a hybrid-integrated chip interferometer for online surface profile measurements *Rev. Sci. Instrum.* **87** 065103

Lasi H, Fettke P, Kemper H, Feld T and Hoffamann M 2015 Industry 4.0 *Buss. Infor. Sys. Eng.* **64** 239–42

Leach R K 2011 *Optical Measurement of Surface Topography* (Berlin: Springer)

Leach R K 2016 Metrology for additive manufacturing *Meas. Control* **49** 132–5

Leach R K *et al* 2018c Information-rich manufacturing metrology ed S Ratchev *Precision Assembly in the Digital Age. IPAS 2018. IFIP Advances in Information and Communication Technology* vol 530 (Berlin: Springer)

Leach R K 2020 *Integrated Metrology 10-year Roadmap for Advanced Manufacturing* (Sheffield: HVM Catapult)

Leach R K, De Groot P and Haitjema H 2018b Infidelity and the calibration of surface topography measuring instruments *Proc. ASPE (Las Vegas, USA)*

Leach R K, Elmadih W, Piano S, Isa M A, Sims-Waterhouse D, Southon S and Syam W P 2020 Development of an all-optical dimensional measuring system *Proc. EUSPEN (Virtual) (June 2020)*

Leach R K, Evans C, He L, Davies A, Duparre A, Henning A, Jones C W and O'Connor D 2015b Open questions in surface topography measurement: a roadmap *Surf. Topogr. Metrol. Prop* **3** 013001

Leach R K, Gisuca C, Guttmann M, Jakobs P J and Rubert P 2015c Development of low-cost material measures for calibration of the metrological characteristics of areal surface texture instruments *Ann. CIRP* **64** 545–8

Leach R K, Giusca C L, Haitjema H, Evans C and Jiang X 2015a Calibration and verification of areal surface texture measuring instruments *Ann. CIRP* **64** 797–813

Leach R K and Haitjema H 2010 Bandwidth characteristics and comparisons of surface texture measuring instruments *Meas. Sci. Technol.* **21** 032001

Leach R K, Jones C W, Sherlock B and Krysinski A 2013 The high dynamic range surface metrology challenge *Proc. ASPE (Saint Paul, USA, April)* pp 149–52

Leach R K, Sims-Waterhouse D, Medeossi F, Savio E, Carmignato S and Su R 2018a Fusion of photogrammetry and coherence scanning interferometry data for all-optical coordinate measurement *Ann. CIRP* **67** 599–602

LeCun Y, Bengio Y and Hinton G 2015 Deep learning *Nature* **521** 436–44

Li D, Cheung C F, Ren M, Zhou L and Zhao X 2014 Autostereoscopy-based three-dimensional on-machine measuring system for micro-structured surfaces *Opt. Express* **22** 25635–50

Li D, Cheung C F, Ren M, Whitehouse D and Zhao X 2015 Disparity pattern-based autostereoscopic 3D metrology system for *in situ* measurement of microstructured surfaces *Opt. Lett.* **40** 22

Li D, Jiang X, Tong Z and Blunt L A 2018b Kinematics error compensation for a surface measurement probe on an ultra-precision turning machine *Micromachines* **9** 34
Li D, Tong Z, Jiang X, Blunt L and Gao F 2018a Calibration of an interferometric on-machine probing system on an ultra-precision turning machine *Measurement* **118** 96–104
Liu Y, Blunt L, Gao F, Jiang X, Zhang Z, Saunby G, Dawes J, Blackham B, Rahman H A and Smith C 2018b *In-situ* areal inspection of powder bed for electron beam fusion AM system based on fringe projection *Proc. ASPE/EUSPEN (Berkeley, USA, July)*
Liu K, Chen Y Q, Zhang T, Tian S and Zhang X 2018 A survey of run-to-run control for batch processes *ISA Trans.* **83** 107–25
Liu M, Senin N, Su R and Leach R K 2020 Cascaded machine learning model for reconstruction of surface topography from light scattering *Proc. SPIE* **11352** 113520Q
Lu P, Chen J and Xie L 2018 Development of decay based PLS model and its economic run-to-run control for semiconductor process *Ann. Am. Control Conf. (ACC) (June, Milwaukee, USA)* pp 27–9
Martin H, Wang K and Jiang X 2008 Vibration compensating beam scanning interferometer for surface measurement *Appl. Opt.* **47** 888–93
Martínez P, Bermudez C, Cadevall C, Ayala A M and Artigas R 2020 Three-dimensional imaging confocal profiler without in-plane scanning *Proc. SPIE* **11352** 113520L
Mendikute A, Leizea I, Yague-Fabra J A and Zatarain M 2018 Self-calibration technique for on-machine spindle-mounted vision systems *Measurement* **113** 71–81
Molleda J, Usamentiaga R, Garcia D F, Bulnes F G, Espina A B D and Smith L N 2013 An improved 3D imaging system for dimensional quality inspection of rolled products in the metal industry *Comput. Ind.* **64** 1186–200
Monti F, Boscaini D, Masci J, Rodola E, Svaboda J and Bronstein M M 2017 Geometric deep learning on graphs and manifolds using mixture model CNNs *Proc. CVPR (Honolulu, Hawaii, July 2017)* pp 5115–24
Moroni G, Petrò S and Syam W P 2014 Four-axis measuring systems performance verification *Ann. CIRP* **63** 485–8
Moroni G, Syam W P and Petrò S 2017 Performance verification of a 4-axis focus variation co-ordinate measuring system *IEEE Trans. Instrum. Meas.* **66** 113–21
Moroni G, Syam W P and Petrò S 2018 A simulation method to estimate task-specific uncertainty in 3D microscopy *Measurement* **122** 402–16
Morse E, Dantan J Y, Anwer N, Soderberg R, Moroni G, Qureshi A, Jiang X and Mathieu L 2018 Tolerancing: managing uncertainty from conceptual design to final product *Ann. CIRP* **67** 695–717
Muhamedsalih H, Blunt L, Martin H, Hamersma I, Elrawemi M and Gao F 2015 An integrated opto-mechanical measurement system for in-process defect measurement on a roll-to-roll process *Proc. LAMDAMAP* XI 99–107
Muhamedsalih H, Jiang X and Gao F 2010 Vibration compensation of wavelength scanning interferometer for in-process surface inspection *10th Proc. Comput. Eng. Annu. Res. Conf.* pp 148–53
Nasrollahi V, Penchev P, Dimov S, Korner L, Leach R K and Kim K 2017 Two-side laser processing method for producing high aspect ratio microholes *J. Micro Nano-manuf.* **5** 041006-1–14
Olesch E and Häusler G 2014 Deflectometry with better accuracy *DGaO Proc.*
Patel R, Hirsch M, Dryburgh P, Pieris D, Yeboah S A, Smith R, Light R, Sharples S, Clare A and Clark M 2018 Imaging material texture of as-deposited selective melted parts using spatially resolved acoustic spectroscopy *Appl. Sci.* **8** 1991

Peharz R, Vergari A, Stelzner K, Molina A, Trapp M, Kersting K and Ghahramani Z 2018 Probabilistic deep learning using random sum-product networks *arXiv*:1806.01910

Pereyra G, Tucker G, Chorowski J, Kaiser L and Hinton G 2017 Regularizing neural networks by penalizing confident output distributions *ICLR Toulon (France, April)*

Quinsat Y and Tournier C 2012 *In situ* non-contact measurements of surface roughness *Precis. Eng.* **36** 97–103

Röttinger C, Faber C, Olesch E, Häusler G, Kurz M and Uhlmann E 2011 Deflectometry for ultra precision machining—measuring without rechucking *Proc. DGaO* **112** P28

Santoso T, Syam W P, Darukumalli S, Cai Y, Helmli F, Luo X and Leach R 2020 On-machine focus variation measurement for micro-scale hybrid surface texture machining *Int. J. Adv. Manuf.* **109** 2353–64

Schleich B, Warmefjord K, Soderberg R and Wartzack S 2018 Geometrical variations management 4.0: towards next generation geometry assurance *Proc. CIRP* **75** 3–10

Schmitt R, Mallmann G, Winands K and Pothen M 2012 Inline process metrology system for the control of laser surface structuring processes *Phys. Proc* **39** 814–22

Schmitt R, Mallmann G, Winands K and Pothen M 2013 Automated process initialization of laser surface structuring processes by inline process metrology *Phys. Proc* **41** 887–95

Schmitt R and Peterek M 2015 Traceable measurements on machine tools—thermal influences on machine tool structure and measurement uncertainty *Proc. CIRP* **33** 576–80

Schwenke H, Knapp W, Haitjema H, Weckenmann A, Schmitt R and Delbressine F 2008 Geometric error measurement and compensation of machines—an update *Ann. CIRP* **57** 660–75

Senin N and Leach R K 2018 Information-rich surface metrology *Proc. CIRP* **75** 19–26

Senin N, Thompson A and Leach R K 2017 Characterisation of the topography of metal additive sureface features with different measurement technologies *Meas. Sci. Technol.* **28** 095003

Sims-Waterhouse D, Isa M, Piano S and Leach R K 2020 Uncertainty model for a traceable stereo-photogrammetry system *Precis. Eng* **63** 1–9

Srivastava N, Hinton G, Krizhevsky A, Sutskever I and Salakhutdinov R 2014 Dropout: a simple way to prevent neural networks from overfitting *J. Mach. Learn. Res.* **15** 1929–58

Stavroulakis P, Chen S, Delorme C, Bointon P, Tzimiropoulos G and Leach R K 2019 Rapid tracking of extrinsic projector parameters in fringe projection using machine learning *Opt. Laser Eng.* **114** 7–14

Stroustrup B 2010 *The C++ Programming Language* (New York: Pearson Education)

Syam W P, Jianwei W, Zhao B, Maskery I, Elmadih W and Leach R K 2018a Design and analysis of strut-based lattice structures for vibration isolation *Precis. Eng.* **52** 494–506

Syam W P, Leach R, Rybalcenko K, Gaio A and Crabtree J 2018b In-process measurement of the surface quality for a novel finishing process for polymer additive manufacturing *Proc. CIRP* **75** 108–13

Syam W P, Rybalcenko K, Gaio A, Crabtree J and Leach R K 2019 Methodology for the development of in-line optical surface measuring instruments with a case study for additive surface finishing *Opt. Laser Eng.* **121** 217–88

Takaya Y 2014 In-process and on-machine measurement of machining accuracy for process and product quality management: a review *Int. J. Autom. Technol.* **8** 4–19

Tosello G 2019 *Product/Process Fingerprint in Micro Manufacturing* (Basel: MDPI)

Tosello G, Hansen H N and Gasparin S 2009 Applications of dimensional micro metrology to the product and process quality control in manufacturing of precision polymer micro components *Ann. CIRP* **58** 467–72

Uhlmann E, Mullany B, Biermann D, Rajurkar K P, Hausotte T and Brinksmeier E 2016 Process chains for high-precision components with micro-scale features *Ann. CIRP* **65** 549–72

Von Rossum G and Drake F L 2011 *An Introduction to Python* (London: Network Theory)

Wang S, Ma J and Gao S 2012 Balancing methods on the three-axis air-bearing platform *Proc. AsiaSim (Berlin, Germany, October)* pp 117–25

Wang Y, Sun Y, Liu Z, Sarma S E, Bronstein M M and Solomon J M 2018 Dynamic graph CNN for learning on point clouds *arXiv*:1801.07829

Weckenmann A and Bernstein J 2013 Optical multi-sensor-measurements in the shop by compensating environmental influences *Proc. CIRP* **10** 61–9

Weckenmann A, Jiang X, Sommer K D, Neuschaefer-Rube U, Seewig J, Shaw L and Estler T 2009 Multisensor data fusion in dimensional metrology *Ann. CIRP* **58** 701–21

Weimer D, Thamer H, Fellmann C, Lütjen M, Thoben K D and Scholz-Reiter B 2014 Towards 100% *in situ* 2D/3D quality inspection of metallic micro components using plenoptic cameras *Proc. CIRP* **17** 847–52

Wilhelm R G, Hocken R and Schwenke H 2001 Task specific uncertainty in coordinate measurement *Ann. CIRP* **50** 553–63

Wollschlaeger M, Sauter T and Jaspernite J 2017 The future of industrial communication: automation networks in the era of the internet of things and industry *IEEE Ind. Electron. Mag.* **11** 17–27

Xu F, Dhokia V, Colegrove P, McAndrew A, Williams H A and Newman S T 2018 Realisation of a multi-sensor framework for process monitoring of the wire arc additive manufacturing in producing Ti-6Al-4V parts *Int. J. Comput. Integr. Manuf.* **31** 785–98

Yamauchi M and Hibino K 2019 Measurement of air turbulence for on-machine interferometry *Appl. Opt.* **42** 6869–76

Yang S M, Jiang X, Maxwell G, Wyatt R and Rogers D 2010 A chip tuneable laser developed for on-line micro–nano scale surface measurements *Meas. Sci. Technol.* **21** 105901

Zangl K, Danzl R, Helmli F and Prantl M 2018 Highly accurate optical μCMM for measurement of micro holes *Proc. CIRP* **75** 397–402

Zhang B, Ziegert J, Farahi F and Davies A 2016 *In situ* surface topography of laser powder bed fusion using fringe projection *Add. Manuf* **12** 100–7

Zhang S 2010 Recent progress on real-time 3D shape measurement using digital fringe projection techniques *Opt. Laser. Eng.* **48** 149–58

Zhang S J, To, Wang S and Zhu Z W S J 2015 A review of surface roughness generation in ultra-precision machining *Int. J. Mach. Tools Manufac* **91** 76–95

www.ingramcontent.com/pod-product-compliance
Ingram Content Group UK Ltd.
Pitfield, Milton Keynes, MK11 3LW, UK
UKHW051853100926
8556IPUK00005B/604

9 780750 325264